Y0-AGJ-354

Last Days of the Concorde

Smithsonian *Air Disasters* Series

Suspenseful stories of tragedy and triumph are brought to life in the Smithsonian *Air Disasters* television and book series through investigative reporting, official reports, and interviews with the pilots, air traffic controllers, and survivors of history's most terrifying crashes. From the cockpit to the cabin, from the control room to the crash scene, the *Air Disasters* series uncovers what went wrong and reveals the changes that were made to ensure such disasters never happen again.

Other titles include:

The Flight 981 Disaster: Tragedy, Treachery, and the Pursuit of Truth

Southern Storm: The Tragedy of Flight 242

SMITHSONIAN

AIR

DISASTERS

SERIES

Last Days of the Concorde

The Crash of Flight 4590 and the End of Supersonic Passenger Travel

Samme Chittum

Smithsonian Books
WASHINGTON, DC

Based on *Air Disasters: Up in Flames*, created by Cineflix and shown in the United States on the Smithsonian Channel.

This book may be purchased for educational, business, or sales promotional use. For information, please write Special Markets Department, Smithsonian Books, P.O. Box 37012, MRC 513, Washington, DC 20013

Published by Smithsonian Books
Director: Carolyn Gleason
Creative Director: Jody Billert
Managing Editor: Christina Wiginton
Editor: Laura Harger
Editorial Assistant: Jaime Schwender
Edited by Gregory McNamee
Designed by Jody Billert
Typeset and indexed by Scribe Inc.

Library of Congress Cataloging-in-Publication Data
Names: Chittum, Samme, author.
Title: Last days of the Concorde : the crash of Flight 4590 and the end of supersonic passenger travel / Samme Chittum.
Other titles: Smithsonian air disasters series.
Description: Washington, DC : Smithsonian Books, [2018] | Series: Smithsonian air disasters series | Includes bibliographical references and index.
Identifiers: LCCN 2018011673 | ISBN 9781588346292 (hardcover)
Subjects: LCSH: Aircraft accidents—Investigation. | Concorde (Jet transports)—Accidents. | Concorde (Jet transports)—History.
Classification: LCC TL553.5 .C486 2018 | DDC 363.12/40944367—dc23
LC record available at https://lccn.loc.gov/2018011673

Manufactured in the United States of America
22 21 20 19 18 5 4 3 2 1

Contents

1

The White Bird

Through its small oval windows—no larger than an outspread human hand—a Concorde jet plane zooming through the stratosphere at supersonic speed offered its passengers a view like no other. The limitless darkness of space seemed almost near enough to touch, while below the distant earth revealed its subtle curve, best appreciated from the passengers' lofty height of 17.7 kilometers (11 mi) above sea level. Travelers were temporary superheroes, flying faster than a speeding bullet, and with a celebratory flute of fine champagne in hand. The futuristic lines of the slender white plane emphasized the rarefied nature of travel within it.

Concorde was a time machine. By reducing intercontinental flight times by almost half, it effectively turned back the clock, making it possible for passengers to leave Paris at 10:30 a.m. and step off the plane in New York at 8:15 a.m. Concorde itself seemed always a step ahead of the present, a hypermodern dream machine whose sleek delta wing, ultrathin fuselage, and characteristic needle nose made it both odd and arresting. Above all, flying Concorde was an experience, one that conferred a feeling of exclusivity and a sense of adventure. Passengers enjoyed the sense of a guaranteed happy ending to such flights, too, for during its 31-year history, not a single Concorde had been involved in a fatal accident.

•

Tuesday, July 25, 2000, was a weekday much like any other at the one of the busiest international airports in Europe, where air traffic controller Gilles Logelin was working the second shift in the southern control tower at the Aéroport Paris–Charles de Gaulle. Early-morning storm activity had delayed flights in France and elsewhere in Europe, creating a backlog of planes waiting to take off and land. Frustrated travelers stood in long lines, only to be told they could not make their connecting flights at other airports. By early afternoon, however, although the air remained muggy, the skies had cleared, making life and working conditions pleasant for controllers stationed near the top of the 22-story tower. "It was a perfect summer day—a typical summer day that you have in the Paris area, and a good day for flying," recalls Logelin. Only a few clouds remained in the otherwise blank blue sky. Summer is holiday and tourist season in France, and July and August are the busiest months of the year at Charles de Gaulle, informally known as the Roissy Airport for the district that surrounds it. Logelin remembers that Tuesday as "a normal day," despite an adjustment to his schedule that at the time seemed trivial. "One of my colleagues had asked to exchange his shift with mine, so I was on duty on this particular afternoon instead of being on duty in the morning."

Four other controllers were stationed in the tower. Logelin was the acting local controller, whose job it was to observe the planes using the two parallel runways that run east to west across the southern sector of the airport near Terminal 2—no small feat, given the 1,500-plus aircraft that landed and took off from Charles de Gaulle every 24 hours. Planes of all sizes, both commercial and private, came and went seven days a week from the gateway airport, which sits 25 kilometers (16 mi) northeast of the capital. Learjets and Cessnas carrying corporate executives to other cities in France made use of the same runways as larger commercial airplanes, including long-range jetliners such as the Boeing 747 and McDonnell Douglas DC-10, and the short/medium-range planes like the Airbus A320 and McDonnell Douglas MD-80.

Thus Logelin's "normal" Tuesday included the usual bustling mass of air traffic, which he monitored both on his radar screen and through the windows of the ultramodern control tower. His perch in the observation post of the concrete and aluminum tower, completed in 1999, was designed for maximum exterior visibility and granted him a unique eagle's-eye view of the two runways to the south. The façade of the tower's supporting column was made primarily of tinted glass; at the very top, the oval observation post where Logelin and the other controllers worked was surrounded by specially made floor-to-ceiling nonreflective windows.

Charles de Gaulle Airport in some ways resembles a small city that must manage and maintain 71 kilometers (44 mi) of taxiways connecting its four runways and three terminals, along with the welter of restaurants, hotels, car rental offices, parking lots, currency exchange counters, and endless outbuildings characteristic of any great airport. In the year 2000 alone, the airport was the pass-through point for 48 million passengers. As befits a country famed for aesthetics, the airport is renowned for its design; it even commissioned its own sleek sans-serif typeface, once called Roissy and now known as Frutiger, to grace its internal signage. Its distinctive 10-story Terminal 1, which opened in 1976 and was designed by international "starchitect" Paul Andreu, is often compared to an octopus because of its modernist circular structure and the iconic white connector tubes that shuttle passengers among its seven satellite gates. Terminals 2 and 3 were added in the 1980s and 1990s to allow the airport to expand. Terminal 2, a series of seven connected buildings, sits south and east of Terminal 1 and is known for the gigantic wood-and-glass-framed hall known as Terminal 2E. The airport has a pair of control towers; the southern one, Logelin's base, sits near Terminal 2.

Charles de Gaulle is the principal hub for Air France, the country's flagship carrier, which has its executive offices on the grounds of the airport, not far from the airport's two southern east-to-west runways. (The airport has four east-to-west runways. Two of the parallel runways lie north of the airport, near Terminal 1, while the other two parallel east-to-west runways are located near Terminal 2 to the south. In the year 2000, however, the airport had only one operational northern runway—one runway was closed for repairs—and two parallel southern runways.)

The airport was laid out to allow personnel in the control towers a close-up view of planes taking off and landing. Each set of two runways includes a longer one for takeoff, closer to the terminals and control towers, and one for landing, farther from the buildings.

Then as now, the south control tower commanded a view of the two southern runways: 26R (26 right) and 26L (26 left). Runway 26R, used for takeoff, is the longer of the two at 4,215 meters (13,829 ft); 26L is 2,700 meters (8,858 ft) in length. "The southern control tower has a very good view over the two runways that we use in the southern sector of the airport," Logelin says. "You can see the entire takeoff of any plane, depending on the weather conditions, of course, and how busy you are. Because you don't only have to watch the aircraft; you also have many things to look at on the radar screen. And it's very busy, always."

Charles de Gaulle dwarfs the commune of Roissy, a small historic village that lies directly to the west. The airport's massive footprint occupies 32 square kilometers (12 sq mi), more than twice that of the village. And its economic footprint is equally broad: the airport employs 100,000 people, including many of the 2,500 inhabitants of Roissy. The village's numerous hotels play host to a steady stream of transient air crews who spend the night there and dine in Roissy's restaurants and cafés. The village acknowledges its huge neighbor with monuments and sculptures commemorating landmarks in the history of French aviation.

Roissy is part of the larger commune, or town, of Gonesse, located about 9.5 kilometers (6 mi) southwest of Charles de Gaulle, and Gonesse residents have long since grown accustomed to the roar of passing planes and the sight of contrails fading away like evaporating kite tails. Many Gonesse residents, too, work at Charles de Gaulle or at Aéroport de Paris–Le Bourget, which caters primarily to privately owned and corporate aircraft and lies directly to the southwest of town. Gonesse has a long history with aviation, beginning in August 1783 when the world's first hydrogen balloon, designed by the French scientist

Jacques Charles and built by the engineers and brothers Anne-Jean and Nicolas-Louis Robert, was launched from the Champ de Mars in Paris. The unmanned red-and-yellow silk balloon traveled aloft for less than an hour before it came to rest in Gonesse, where frightened peasants, thinking it was a monster, attacked it with knives and pitchforks.

•

Among the passengers streaming into crowded Charles de Gaulle on the sunny afternoon of July 25, 2000, were 97 men and women and three children who held tickets for a rare and special flight: Air France Flight 4590, departing at 3:25 p.m. for John F. Kennedy International Airport in Queens, New York, aboard a chartered Aérospatiale/BAC Concorde. Most of the passengers were German holidaymakers, and they were in a festive mood as they prepared to embark on an exotic excursion that would take them first to Manhattan, where they would board a cruise ship, the MS *Deutschland*, and sail at a leisurely 20 knots (23 mph) to Manta, Ecuador, via Florida, the Bahamas, and the Panama Canal. Their ambitious itinerary was arranged by Peter Deilmann Cruises, a tour company based in Neustadt, Germany, and the ticketholders had paid $10,000 each for their extended holiday.

The MS *Deutschland* was a luxury liner with Roaring '20s–themed décor. It was only a third of the size of the *Queen Elizabeth 2*, but it was decorated with more than a nod to old-fashioned glamour and the grand days of cruising: It held a saltwater pool, a ballroom with a glittering chandelier, an Art Deco–styled cinema, and a café where passengers perched on cast-iron chairs beneath globe-shaped streetlight lamps that "could have come straight off Berlin's Unter den Linden boulevard," according to one travel writer. Launched in 1998, the cruise ship was only two years old, and all the trappings—marble pillars, brass fixtures, and ivory-painted panels—were in pristine condition.

Like all of its French and British siblings, Concorde 203, as it was numbered by its manufacturer, was aesthetically arresting but far from brand-new. Registered as F-BTSC, the chartered Concorde being readied that Tuesday had taken its maiden flight in 1975 but was not

Concorde 203, registered as F-BTSC, at Aéroport Paris–Charles de Gaulle in July 1985. One of the six Concordes in Air France's fleet, it would crash as Flight 4590 in Gonesse, France, on the afternoon of July 25, 2000. (© MICHEL GILLIAND)

purchased by Air France until October 23, 1980. Still, it had logged only a modest 11,989 flight hours since it began commercial service 20 years earlier, fewer than 1,000 hours per year. By contrast, a typical workhorse Boeing 747 flies 3,000 or more hours over the course of a single year.

The manufacturer determines the projected lifespan of any aircraft. The true *age* of any airplane, however, is measured not in years but in the wear and tear produced by flights or cycles—combined takeoffs and landings that put stress on the plane's airframe. A plane required to make many short-haul flights will experience more metal fatigue and therefore will age faster. Concorde was built for long transatlantic crossings such as the one from Paris to New York, a distance of 3,153 nautical miles (5,840 km or 3,629 standard mi). Still, supersonic planes are subject to unique stressors, including high surface temperatures on their wings, fuselage, and fin as they blaze along at

2,172 kilometers per hour (1,350 mph). Just the day before, on Monday, July 24, British Airways had reported the discovery of tiny cracks in the spars (crossbeams) of the wings of all seven of their Concordes, but the airline insisted that the cracks posed no safety risks. Concorde 203, one of six operated by Air France, had undergone routine maintenance at Charles de Gaulle Airport four days before, on Friday, July 21, when it was given the equivalent of a doctor's clean bill of health.

Concorde 203 had arrived late Monday evening at Charles de Gaulle following an uneventful flight from New York. As usual, it had been towed to a maintenance bay the following morning for a routine check of all four of its engines carried out by Air France mechanics. Around 3 p.m., the plane was towed to a parking bay near the southern control tower. There it awaited the flight crew, whose members would perform their own series of preflight checks before takeoff.

A one-way ticket from Paris to New York aboard Concorde was then selling for the eye-watering sum of £7,000 (€7,877 or $9,150) and up, priced not only for the massive cost of the fuel required to propel the jet at supersonic speeds but also for the sheer exclusivity of its passenger rosters. Those passengers typically included members of the wealthy, high-flying elite of Europe and North America. "They were, what we called in the '60s, members of the jet set," says former Concorde captain Jean-Louis Chatelain. "We had really unique passengers, from top chefs to people from Hollywood and the world of fashion. And we also had many scientists on board, and Nobel Prize winners over the years." For such passengers, the costly ticket was fair payment for the ease and prestige of flying on the sleek supersonic passenger plane, which whisked them from New York to Paris or London in less than three and a half hours, as compared to the usual eight hours required to make the same trip on an ordinary subsonic airliner. Passengers flying British Airways roundtrip from London to New York could take a Concorde out of Heathrow Airport in the morning and be home again in London that same evening.

On this particular Tuesday, however, the most famous occupant of Flight 4590 would not be a passenger but its captain, Christian

Marty, 54, who had achieved celebrity status in France for his daring exploits as a sportsman. Nicknamed Kinou, he was not only a member of the rarefied club of Concorde pilots but also a famed windsurfer, rally driver, skier, cyclist, and hang-glider. He had once flown his glider over a volcano and was renowned for his February 1982 windsurfing odyssey across the Atlantic, a 5,000-kilometer (3,107-mi), 37-day trip from Senegal to French Guiana. It was a perilous undertaking during which Marty slept tied to his board at night. He nearly lost his life one night when he fell off his windsurfer, which was carried away in the dark by ocean currents, forcing Marty to swim to retrieve it.

The honor of having a celebrity fly their plane would have seemed entirely appropriate to those looking forward to their flight on Concorde. (The plane itself was a minor celebrity: Concorde F-BTSC had one brief star turn in 1979 in the American-made action movie *Concorde, Airport 79*, in which it eludes a cruise missile fired by terrorists.) Among the privileges accorded them were access to a private lounge and the chance to chat with Kinou himself before departure. "As the captain, you would go to the business lounge," recalls Chatelain. "And in the business lounge there was a first-class lounge. And in the first-class lounge there was a specific room dedicated to Concorde passengers. We would go there and meet the passengers before the flight. So it was a kind of relationship between the crew and passengers that does not exist anymore."

Most of the passengers preparing to board that day were over 60 years old, but two families were traveling with children and grandchildren. Kurt Kahle, 51, who ran a private business school in Mönchengladbach, Germany, and his wife, Marion, 37, were accompanied by their son, eight-year-old Michael. Andreas and Maria Schranner, 64 and 62, were wealthy Munich property magnates spending their adventure-filled retirement traveling and taking cruises. Today they had brought along their daughter, Andrea, 38; her husband, Christian Eich, 57; and Andrea's two children, 10-year-old Maximilian and eight-year-old Katharina. Christian ran a museum for BMW, and Andreas looked forward to celebrating his 65th birthday on the trip. Another retired couple was waiting to board that day as well: Rolf and Doris Maldry, 68 and 64, retired civil servants from Schwerin, Germany, who had saved for years to take a dream vacation that included a flight on Concorde, a travel experience like no other.

The Eich children had been looking forward to their glamorous family holiday and had talked of little else in conversations with friends and classmates during the weeks leading up to the two-week trip. Maximilian and Katharina's unbridled excitement was equaled by that of an adult passenger who later would be described in the press as a "Concorde fanatic." Klaus Frentzem, 53, was a secondary schoolteacher who had long dreamed of taking a trip on Concorde and who for 20 years had patiently set aside money from his modest salary in order to afford the price of the ticket. He and his wife, Margaret Frentzem, were flying together, taking the honeymoon they been unable to afford as a young couple. They were two among a group of 13 hailing from Mönchengladbach, a town near Germany's border with the Netherlands. As Frentzem waited in the departure lounge, he showed off his collection of Concorde memorabilia, including toy airplanes, which he had brought along for the trip. Also preparing to board Concorde that day were well-known society photographer Irene Vogt-Götz and her husband, Christian Götz, an author and former trade unionist. Residents of Düsseldorf, both were in remission after battling cancer—she had been diagnosed first in early 1999 with breast cancer, while Christian was diagnosed with esophageal cancer later that same year. The flight to New York and the cruise to follow were to be a celebration of their future together and their shared status as cancer survivors.

It would be a full flight. Concorde could seat 40 in 10 rows in its forward cabin, plus 60 passengers in 15 rows in the larger aft cabin. A single aisle separated the two-abreast seats on each side. (The seats were numbered 1 to 26, since row number 13 had been omitted, a common practice given the number's association with bad luck.) With its low ceilings and narrow fuselage, Concorde was neither spacious nor especially comfortable. Even so, Concorde passengers could drink champagne and dine on lobster canapés while they gazed out their small, double-glazed windows at a stunning panorama that included the very curvature of the earth and, high above, the darkening indigo of outer space. They could watch the plane's accelerating Mach number on a digital readout in the cabin, and the captain would announce the precise instant the plane broke the speed of sound. *Telegraph* travel editor Graham Boynton once wrote about the experience in a personal tribute to Concorde. "This is where astronauts and fighter pilots fly, only they're in spacesuits and you're in chinos and a cotton shirt, sipping fine wine and listening to Mozart on headphones. From Concorde I have seen the Northern Lights, the deep blue of outer space, the curve of the Earth, and have travelled faster than most of the human race can even imagine."

•

Concorde was at that time the only supersonic commercial passenger jet in the world. Its lone rival, the failure-prone Soviet-era Tupolev Tu-144, had been demoted by Aeroflot to the far less glamorous role of cargo plane following two high-profile crashes in 1973 and 1978, the first at the Paris Air Show and the second during a test flight.

Only 20 Concordes had ever been built, including prototypes and preproduction models, and just 14 were destined to enter commercial service for British Airways and Air France; each airline purchased seven Concordes. In 2000, only six of the original seven Air France Concordes were being actively flown after the airline retired one in 1982 to be used as a ready source of spare parts for the remaining six. Concorde's safety record was enviable: Not a single life had been lost in a crash in its 24 years of commercial service, inaugurated with regularly scheduled flights that began in 1976.

The brainchild of an unusual collaboration between the British and French governments, the supersonic Concorde was arguably the epitome of technically ambitious thinking, the product of an optimistic era when it was hoped that a supersonic plane would revolutionize the rapidly expanding air travel market. Because of low oil prices, fuel consumption was not yet an issue when Concorde was conceived in the

1960s. Concorde was a gas guzzler that used 4.3 US gallons of fuel to go just 1 kilometer. A Concorde required four times as much fuel per passenger to wing its way across the Atlantic than its more fuel-efficient rival, the Boeing 747.

There had been only one generation of Concordes, all built in the 1970s and early '80s. Concordes had been crossing the Atlantic since 1976, when two flew inaugural flights in tandem from London to Bahrain and Paris to Rio de Janeiro. In addition to being fuel-thirsty, Concorde was astonishingly noisy, and most airlines had refused to buy them, fearing their thunderous sonic booms—*les bangs soniques*, in French—would make them unacceptable for flying profitable cross-country domestic routes among major cities in the United States and Europe. Yet Concorde retained its cachet as a superior plane and remained in demand for transatlantic and charter flights to destinations around the world, including a July 1985 charter flight for 60 children who won prizes that earned them a free, once-in-a-lifetime trip to Legoland Billund, the original Lego theme park in Denmark.

With appearances on the runways of airports in New York, London, and Paris, Concorde never lost its allure with the passage of time. "It was kind of a show," recalls Chatelain. "I mean, all pilots starting to taxi their aircraft would make sure that they were in a good position to watch Concorde at takeoff or landing. I could see that on each flight. Reaching Kennedy airport in New York, we could see that those American pilots, they were also looking at us at landing or takeoff."

The single most eye-catching feature of Concorde is its swept-back delta wing, so named for the triangle-shaped fourth letter of the Greek alphabet. Seen from above, the aircraft's total shape is that of a slender triangle, the cockpit at its apex. The basic shape of a delta-wing aircraft is much like that of the familiar child's folded paper airplane, but its simple, bold lines are well suited to the rigors of supersonic flight. It is no coincidence that both the Soviet and Concorde supersonic transport (SST) projects chose the delta-wing design. As it moves through the air at speeds above Mach 1, the nose of a supersonic aircraft produces a V-shaped shock wave in the air, similar in shape to the bow wave that a boat pushes forward in water. Ordinary planes produce a pressure wave in the air as well, of course, and when the air impacts the wings of a standard plane, the drag forces that result impede forward motion and keep the plane at subsonic speeds. The wider the wingspan, the greater the drag forces; the narrow, thin, short-span delta wing of a supersonic aircraft, however, fits neatly inside the shock wave and thus reduces drag.

Concorde was far more than the sum of its parts, however. Aviation enthusiasts who had flown on a Concorde or seen one take off typically struggled to find words to describe the effect it had on them. "When it's departing, it's a magical thing," says Bob MacIntosh, a retired U.S. Air Force pilot and air crash investigator for the National Transportation Safety Board. "You see the white fuselage and the droop nose. It rotates extremely high in comparison to other aircraft, and then, as it passes, you feel it and you hear it, because it is loud and strong. And it's just a beautiful sight to both the spectator and the aviator."

Among its admirers was Logelin, who never tired of watching the supersonic jet as it prepared for takeoff from Charles de Gaulle, where his post in the southern control tower afforded him an unobstructed 360-degree view of dozens of planes coming and going at a rate of more than 148 movements, or takeoffs and landings, per hour. "Every plane is nice, but Concorde was different," observes Logelin. "It had a kind of a sharp shape, and maybe because of its color—it was usually white—it was like a nice bird that you are looking at while you work. It's more like a fighter jet with its delta wings, which give the idea of speed and also of power and efficiency."

Although it was often compared to a swan, Concorde had two storklike feature: its long main landing gear legs. The single forward landing gear with two small tires was located in the front and middle of the fuselage, supporting the nose of the plane. Most of Concorde's weight was borne by its two main left- and right-side landing gear legs and bogies, or sets of wheels, each fitted with four heavy-duty nylon bias-ply tires. The two long legs, located under the wing, were retractable and hydraulically controlled, allowing the pilots to stow the legs and bogies in a well or gear housing inside the fuselage after takeoff.

Logelin and the other air traffic controllers would typically allow their gaze to linger on the sleek jet as it began its maneuvers in preparation for takeoff. "If the weather is good, you could see the entire takeoff of any plane, depending on the weather condition, of course." It was perhaps excitement as much as the opportunity for a better view that would prompt Logelin to rise to his feet as the supersonic jet roared down the runway prior to its dramatic liftoff, powered by four Rolls-Royce/SNECMA Olympus 593 turbojet engines, each with 169,032 Newtons (38,050 pounds) of thrust.

Logelin had seen Concordes in performance mode numerous times in the past, but that fact only added to his anticipation. The planes' thundering, high-speed takeoffs were always stimulating high points in his busy but routine workdays. July 25, though, would be anything but routine.

2

"You Have Flames behind You"

By 4 p.m., Captain Christian Marty, First Officer Jean Marcot, 50, and Flight Engineer Gilles Jardinaud, 58, were at work in the cockpit, running through their preflight checklist while Concorde F-BTSC sat parked near Terminal 2. Outside, two Air France mechanics had brought out tools to make a last-minute repair.

Concorde's four power plants are mounted beneath the slender delta wings in housed pairs, numbers 1 and 2 on the left and 3 and 4 on the right. A small operational flaw in the performance of a pneumatic motor on the number 2 engine's thrust reverser had been observed and logged by the flight engineer, one of the crew who had made the round-trip from Paris to New York and back the day before, Monday, July 24.

The thrust reverser helps slow down the aircraft during landings, but its absence would not compromise the plane's safety, and manufacturer guidelines allowed the plane to fly without the repair being made. Yet Marty insisted the defective part had to be fixed before takeoff. Concorde F-BTSC would carry a full complement of 100 passengers, plus a full crew and its usual full load of fuel, and Marty wanted everything in order for a problem-free landing in New York.

Locating a new replacement motor proved impossible, so another Concorde parked at the airport was cannibalized; the Air France mechanics installed the purloined motor and finished the repair in 30 minutes. The problem was resolved, but it was one more hang-up in a day fraught with delays; the early afternoon flight from Frankfurt, carrying German passengers with tickets to fly on Concorde, had arrived late, followed by problems in transferring their luggage. It was soon apparent that Flight 4590 would not be leaving Paris at the scheduled 3:25 p.m. departure time, which came and went while the plane sat parked at the gate, waiting for the repair to be completed and all the luggage to be counted and loaded.

Still, the passengers remained in high spirits. Some were even heard singing in the departure lounge before they boarded. By 4:12 p.m., all were in their seats, and the cockpit crew was running through its next routine task, the flight-deck review of fuel and oil, switches, filters, and circuit breakers. Like all commercial pilots, the crew set the so-called V speeds, shorthand drawn from the French word *vitesse*, or "speed." V speeds are standard, prescribed safe speeds for the various phases of takeoff and flight. They differ from one model of plane to another and can be adjusted by pilots prior to takeoff as they consider runway conditions and other factors, including the weight of cargo and fuel on board.

The three most critical speeds during takeoff are known as V1, VR, and V2min. V1 is the decision speed, when the aircraft can no longer stop to abort a takeoff. VR is the speed at which the pilot pulls back the yoke to raise the nose of the plane into the air. V2min is the safety takeoff speed—the speed that an aircraft suffering from the failure of an engine must reach to allow the plane to achieve the necessary climb gradient or height off the ground at the end of the runway. That day, the takeoff decision speed, or V1, was set at 150 knots (172 mph). The VR and V2min selected were 198 knots (227 mph) and 220 knots (253 mph), respectively.

Marty, slated to pilot the plane at takeoff, had been a Concorde captain since qualifying in August 1999, and he had had his last pilot's competency check in June 2000. He had accumulated 13,477 flying hours (317 on Concorde), including 5,495 as a captain. Before that, he had flown the Airbus A340 and trained other pilots as an instructor. Sitting in Concorde's right-hand seat was Marcot, a highly experienced first officer who had been flying Concordes since 1989. He was also an instructor on the Air France Concorde simulator and had 10,035 flying hours, of which 2,698 were spent as first officer on Concorde. First Engineer Jardinaud had a little more than three years' experience on Concordes and 12,532 total flying hours, 937 of them as a member of a Concorde flight crew.

There were 95 tonnes of fuel in the tanks. The plane, as was normal, would use up a ton of jet fuel just taxiing out to get in position for takeoff. The flight crew made allowance for that as well. Every Concorde was made heavy by the large supply of fuel stored in 13 tanks set in its fuselage, wings, and tail. Managing the fuel was one of the trickiest parts of flying the plane. The distribution and total weight of the fuel determined Concorde's center of gravity, which had to be adjusted as the supply of fuel diminished. Pilots also had to adjust for the weight and location of the passengers and cargo. Only once they had everything balanced out to maintain the proper center of gravity could the captain initiate the takeoff procedure. If they took off with the center of gravity too far forward or backward, the stability of the plane could be compromised. Once in the air, maintaining the center of gravity was vital to keeping the aircraft level, so the flight engineer continuously monitored the that moved the fuel back and forth among the tanks (see page 72).

At 4:20 p.m., the Air France dispatcher, who helped monitor the weight of the baggage and cargo being loaded, came into the cockpit to inform the crew that they were carrying 2.2 tons of baggage, prompting them to update the takeoff weight to 185.07 metric tons. As the moment for departure drew near, it was time to rev up the plane's four powerful engines: 1 and 2 on the port side and 3 and 4 on the starboard side. Even idling, the plane's engines were formidable. Startup procedure called for powering up two engines at a time: one engine on each side of the plane to keep the aircraft balanced and stable. So at 4:25 p.m., engines 3 and 2 (inner engines on the right and left) were started up, to be followed a few minutes later by engines 4 and 1 (outer engines on the right and left). Meanwhile, Marty and Marcot monitored the radio frequency used by the air traffic controllers. At 4:34 p.m., Gilles Logelin cleared the aircraft to taxi to runway 26R via the Romeo taxiway. "At a certain point," Logelin says, "the aircraft is approaching what we call the holding point—the point where it could enter the runway or not, depending on the authorization he's getting. Of course, whenever a plane seeks to enter a runway it's something very important. The aircraft needs a clearance for doing that. So the first contact [with Concorde's captain] was to let him line up on the runway and prepare for takeoff."

Because Charles de Gaulle Airport is used by planes of all sizes, it is the controller's job to allow enough time for smaller, slower planes to depart safely before giving the go-ahead to aircraft that are much speedier on the ground. "There is what we call the departure sequence, and it's always a mixing of aircraft departing and arriving," says Logelin. "I had to wait before letting Concorde take off in order to make sure that it was not going to gain on smaller aircraft." Although he had seen Concordes in action before, Logelin felt a sense of anticipation as he saw the sleek and powerful jetliner, with Captain Marty acting as pilot, take its position as next in line to enter the runway for takeoff. "Concorde has a very spectacular departure. When Concorde is taking off, it's very graceful. The lifting up is very interesting to see."

•

Concorde had a much higher angle of attack for takeoffs and landings than a regular plane did. Its needle nose pitched upward at a 14-degree angle, much steeper than the typical range of angles for subsonic planes. Another reason a Concorde takeoff was such a spectacle was the earth-shaking roar of the engines, reminiscent of a rocket blasting off a launchpad. The noise was the result of combustion within the jet's four huge Rolls-Royce/SNECMA Olympus 593 engines and the engines' afterburners, which supplied the additional, fuel-injected burst of power required to get the plane airborne. The roar of the afterburners kicking in could be both heard and felt. "When I was in the office at Charles de Gaulle," says Jean-Louis Chatelain, "I could tell you when Concorde was at takeoff because of the windows vibrating."

Concorde needed to sprint much faster and farther down the runway than a subsonic jet to achieve the speed needed to get the craft off the ground. To go airborne, Concorde had to reach speeds approaching or surpassing 220 knots (250 mph), as opposed to the 130 to 155 knots (150 to 173 mph) that most subsonic jets must reach. A subsonic plane is aided during takeoff by the lift created by its wing flaps, which Concorde lacked. Instead, Concorde depended largely upon the brute force of its engine power to lift itself off the ground. Once airborne, Concorde pilots had to accelerate as quickly as possible to reach supersonic speed, where their planes were designed to fly efficiently. But, having achieved a safe climbing speed after takeoff, the crew had to cancel the reheat to reduce noise and remain subsonic until they were over the sea and then relight it to accelerate to supersonic speed.

At 4:42 p.m., Logelin radioed the wind speed to the Concorde crew and cleared the aircraft for takeoff, providing vital information about localized surface wind direction and speed: "Air France four-five-nine-zero, runway twenty-six right, wind zero ninety [degrees], eight knots. Cleared for takeoff." Marty did not respond to this information. Light breezes averaging 3 knots prevailed that afternoon in the area of runways 26R and 26L. Logelin was effectively cautioning Marty by letting him know that a more robust east-to-west breeze had arrived. Neither the direction nor the speed of surface winds reported by Logelin was ideal for takeoff, and both factors would later be sources of controversy, as was Marty's silence in response to Logelin's report.

Instead of feedback from Marty about the problematic tailwind, Logelin heard First Officer Marcot's confirmation that the flight crew was ready to go: "Four-five-nine-zero. Take off twenty-six right."

The next voice was that of Captain Marty: "Is everybody ready?"

Marcot replied with a businesslike "Yes." Flight Engineer Jardinaud followed up: "Yes."

As Flight 4590 entered runway 26R, Logelin was on his feet, as usual, to get an optimal view as the plane unleashed the power of its engines and began accelerating down the long takeoff runway. "I had been watching the plane quite a lot up to that point. And as it was the

Concorde, the temptation was great to observe it taking off." For the first 1,800 meters (1.12 mi) from the threshold of the runway, all looked well. But just seconds later, at 4:43:13 p.m., Logelin saw something startling and unprecedented in his experience as a controller: bright yellow-and-red flames streaming directly beneath the underside of the port wing from the area of the landing gear toward the port-side engines at the trailing edge of the wing to the tail and beyond—creating a fiery tail about 61 meters (200 ft) long as the plane lifted off the runway. The plane was on fire. "I never experienced that before—to see flames on an aircraft that is departing on the runway," he recalls.

Immediately, at 4:43:13 p.m., Logelin was on the radio to alert the Concorde flight crew that their plane was on fire, saying: "Concorde zero, four-five-nine-zero, you have flames, you have flames behind you."

The takeoff could not be aborted. At the moment Logelin first spotted the angry flames, the plane had already lifted up its single front landing gear as Marty began rotation, while the twin rear main landing gear and wheels remained on the runway. "I knew it was the sign that the pilot had already overcome the [V1] speed, where the plane cannot do anything else other than take off. Immediately I pushed the red button, which is the button for [emergency] alert," Logelin says.

The vital purpose of the alarm, observes Logelin, is to alert all the emergency services that must respond in the event of an accident, including fire services, the airport police, "and of course the main control room, the approach room." At the same time that a loud alarm bell was ringing in the airport's fire services garage, other controllers who were stationed in the control approach room, located 2 kilometers (1.2 mi) away in the northern control tower, heard an electronic buzzer, installed "so that everyone knows that something is going on."

As the alarm went out from one end of the airport to the other, Logelin received a one-word response from Captain Marty. "'*Roger*': the pilot just answered me with the typical word in aviation," Logelin recalls. "It means 'OK, I understand.' And that's all he said. Because I'm a pilot myself, I knew they were busy. So the fact that they acknowledged my message was enough for me: OK, they know. Now they know they have a problem."

It wasn't long before he saw several fire and rescue vehicles, lights flashing, racing behind Concorde as it continued to hurtle onward, spouting flames that were growing in size and intensity. "I sent another radio message a short time after that to advise them that the flames were gaining in size. It was difficult to say any more than that. If I had seen the flames becoming shorter, it would have been a good indication that the situation was maybe under control. But I could see the flames were bigger, which was not good."

The atmosphere in the control tower was "very tense" as Logelin worked as quickly as possible to divert traffic and clear the runway in case Concorde was able to return to the airport for an emergency landing. "It was important to me to keep calm to be able to control the rest of the aircraft and make all the decisions that I had to make to clear the runways and direct other aircraft to go around. Everything was done in collaboration with the northern tower, because we had to send aircraft to other runways. So it was a very, very busy time, not only for me but for all my colleagues. I was in a tense situation, and they were, too." Left unspoken but understood was the possibility that the plane's tanks could explode. Such a catastrophic event would doom everyone on the plane. It would also endanger more people on the ground if Concorde were near a building.

Logelin and the other controllers were schooled in the airport's protocol for dealing with emergencies. They had been thoroughly trained in how to coordinate their response with those of fire and rescue personnel, and Logelin had put that training to use during minor emergencies in the past. Once the pilot of a plane approaching Charles de Gaulle had suffered a heart attack, and Logelin had quickly called for an ambulance to meet the plane on the tarmac while he redirected other incoming aircraft so the copilot could take over and make an emergency landing. But the Concorde crisis was a thousand times worse, and it was unfolding with shocking speed.

"Things were going very fast. I was watching the plane as best I could, but not always, because my eyes were moving back and forth from the plane to the radar screen. I wanted to make sure if they wanted to come back on my runway, that they could. I needed to clear the traffic and clear the runway." His efforts required all his concentration. There was no need for him to ask questions or request an update from the pilots of the stricken plane. "In this kind of a situation, you cannot interfere with the cockpit's work. My job was not to ask them anything because what I could see and what I knew were enough."

At that point, Logelin had only a very general idea of what mechanical and logistical problems the flight crew was trying to overcome. The first officer had not sent any further transmissions or made any requests for assistance. "I didn't know which option the crew would choose. I didn't imagine that they could not land." Nor did it occur to him "that everything was lost. It was my duty to make it possible for them land somewhere. I didn't expect anything from the flight crew. I was open to everything. I was prepared to clear the traffic, to clear the sky, to do whatever I could if it could help. And obviously, I could not know in advance what was going to be the pilot's decision because I was not in the cockpit. All I knew was that the situation in the cockpit was very, very tense."

As a pilot himself, Logelin was aware that Marty, Marcot, and Jardinaud were furiously troubleshooting and making rapid-fire decisions about what, if anything, could be done to contain the fire, control the plane, and get it back on the ground as soon as possible. "I knew that they were doing their best to save the plane if they could. They're deciding in real time. And I knew also that I would do the same on my end. We are all professionals. So it means we don't interfere in each other's work, especially in this kind of emergency where the situation is very, very bad." In France, many air traffic controllers are also pilots, "and that gives us a very good understanding of what can be the situation in a cockpit in case of emergency. This helps a lot to deal with any emergency."

Nonetheless, while Logelin worked methodically to divert planes and clear a runway for Concorde F-BTSC, he could hear on his open frequency a series of frantic, staccato commands and urgent reports from the cockpit, where the plane's engine fire alarm had gone off and could be heard blaring in the background.

He could also hear a report over the radio from an incredulous observer, an unidentified pilot, who could also see that the plane was on fire: "It's really burning, eh?"

At 4:43:25 p.m., Captain Marty's voice was heard: "Engine fire procedure."

Two seconds later, Logelin heard the clearly alarmed First Officer Marcot shout a warning: "Watch the airspeed, the airspeed, the airspeed!"—then only about 200 knots (230 mph), far slower than it needed to be to gain altitude.

"Gear on retract," Marty said at 4:43:30 p.m., hoping to pick up speed by using controls that initiated retraction of the plane's still fully extended landing gear legs and bogies.

At the same time, Logelin could hear the same unidentified pilot make another report on what he was witnessing: "It's really burning, and I'm not sure it's coming from the engines."

Just moments later, Logelin was back on the radio. "Four-five-nine-zero, you have *strong* flames behind you." As he informed the crew that the fire engulfing the back of the aircraft was getting worse, he could hear First Officer Marcot struggling to get control of the unresponsive landing gear as a second fire alarm went off.

At 4:43:45 p.m.—32 seconds after Logelin had alerted the crew to the fire—First Officer Marcot's voice could be heard once again. "I'm trying," he said as he struggled with the controls for the landing gear legs, which remained fully extended, creating tremendous drag on the plane.

Then a question from Captain Marty for Jardinaud: "Are you shutting down engine two there?"

Two seconds later, Jardinaud replied: "I've shut it down."

One second later, First Officer Marcot was heard once more invoking the airspeed and then making a further report on the landing gear: "The gear isn't retracting." This declaration was nearly drowned out by the blare of the fire alarm, which resumed ringing after being shut down. Less than a minute had passed since the crisis began, and three ground-proximity warning systems had gone off, emphasizing what the pilots already knew: the plane was in immediate danger of hitting the ground. Marcot again announced, "The airspeed!"

At 4:44:05 p.m., Logelin directed the airport's fire chief to be prepared to move tanker trucks into position if the Concorde managed to return to the airport and land on one of the southern runways: "Fire service leader, err, the Concorde . . . I don't know his intentions. Get into position near the southern double runway." Anticipating a different landing option, Logelin changed his mind and directed the fire services to another runway at the northern end of the airport—the longer runway normally used for takeoffs. "Fire service leader, correction. The Concorde is returning on runway zero nine in the opposite direction."

But First Officer Marcot had another last-ditch hope while Concorde, still on fire, was losing thrust and maneuverability as it continued flying west-southwest toward Gonesse and away from Charles de Gaulle, with no hope of circling back. Directly south of the plane was Le Bourget Airport, which was much closer than Charles de Gaulle. If the plane kept going, it might make it to one of that airport's three runways.

"Le Bourget," Marcot called out at 4:44:14 p.m., and yet again, "Le Bourget." Eight seconds later came his response to Logelin's efforts to clear a northern runway for an emergency landing at Charles de Gaulle: "Negative. We're trying for Le Bourget!"

It was his last recorded sentence. At 4:44:31 p.m., less than two minutes after Logelin called Concorde to report a fire engulfing the rear of the aircraft, the cockpit voice recorder stopped working.

Logelin did not require an explanation as to why the Concorde pilots were opting to attempt to put the plane down at Le Bourget Airport to the south and west of Charles de Gaulle. "Le Bourget is very close to Charles de Gaulle," he says. "So if the situation of the plane was so bad that the pilot thought that he could do nothing to land except near where he was, it made sense to land on a runway that is nearby. He

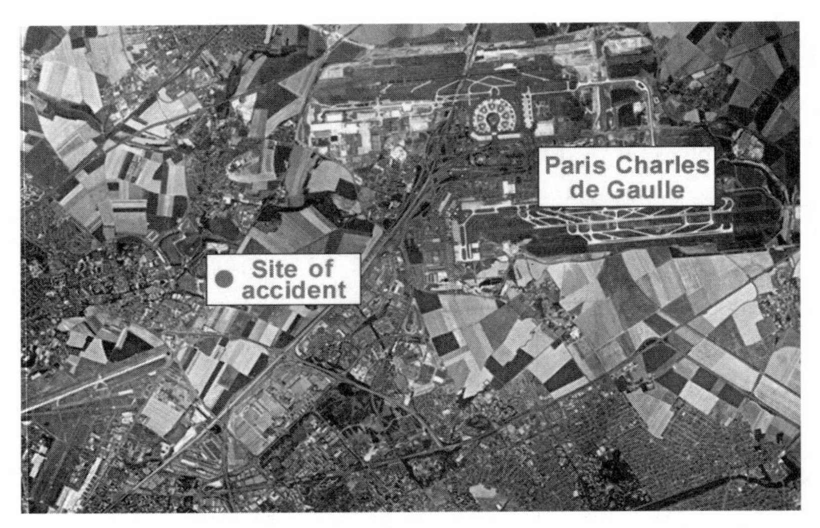

Map of Concorde's brief flight: Charles de Gaulle Airport is at the upper right, with its distinctive octopus-shaped Terminal 1. Flight 4590 used the east-west runway 26R for takeoff. To the southwest is the A1 Autoroute and the crash site: the Hotelissimo in Gonesse. At the far southwest corner of the photo is Aéroport de Paris–Le Bourget, which First Officer Jean Marcot indicated the flight crew were trying to reach just before Concorde crashed. The flight covered only about 9.5 kilometers (6 mi). (BUREAU D'ENQUÊTES ET D'ANALYSES POUR LA SECURITÉ DE L'AVIATION CIVILE, FRANCE, *ACCIDENT ON JULY 25, 2000, AT LA PATTIE D'OIE IN GONESSE [95] TO THE CONCORDE REGISTERED F-BTSC OPERATED BY AIR FRANCE*, REPORT TRANSLATION F-SC000725A, ENGLISH ED. [JANUARY 2002], 67)

could not do anything else. . . . Marcot said, in effect, 'OK, I will try Le Bourget, there was no time to do anything else.' So he tried. That's all. It was an instant decision. My job was to let the pilot do whatever he could to save the plane and the passengers."

Only one minute and 18 seconds had elapsed between 4:43:13 p.m., when Logelin spotted the fire, and 4:44:31 p.m., when the recording of the words and sounds in the cockpit came to an abrupt end. During that time, Captain Marty, First Officer Marcot, and Flight Engineer Jardinaud had never stopped trying to control their crippled, fire-ravaged plane and trying to avoid a catastrophe. In the end, they ran out of time. Their courage had not failed them, but their aircraft could not recover from the spreading fire, landing gear malfunctions, and fatal damage to the engines. After losing speed and altitude at an accelerating rate, Concorde had traveled only about 9.5 kilometers (about 6 mi) from Charles de Gaulle before it plummeted to the ground in Gonesse.

•

"At the very moment when the aircraft crashed, at the split second it happened, my eyes were on the radar," says Logelin. "So I didn't see the action of crashing." But instead of the shouted commands of the first officer, Logelin now heard another voice, that of a pilot who had seen the entire drama play out and yet could barely believe what he had witnessed. "Also on the radio there was an aircraft using a blind transmission," says Logelin. "And a pilot somewhere said, 'Oh, my God, that's it! It's on the ground.' So that's when I realized the aircraft had crashed. And it was hard to believe. But . . . I could see kind of a big mushroom of smoke, which was the sign that the plane was on the ground somewhere."

Logelin's job was not done. He had another grim but essential task to perform. "When I saw this mushroom cloud, I called the Air France plane. Twice I called: 'Air France 4-5-9-0, do you hear me?' But there was no answer." The silence on the other end and the dark, billowing smoke filling the sky to the southwest told him that the hard-fought battle to save the plane was over, a reality he found difficult to accept. "Until the very last moment, I was thinking that something would save the situation. I never realized that they were not going to succeed." To do his job, he says, "I have to believe in what I am doing, the same as the crew and captain. They have to believe that they are going to succeed. *I* had to believe they were going to succeed."

Nonetheless, Logelin had to press on and deal with the immediate aftermath of the crash as every available rescue vehicle rushed to the scene. "I gave instructions to keep all the aircraft on the ground and to cancel all departures. And apart from a few aircraft that landed this day, the others had to go around and go to the other runway. There was an Airbus on approach just a few miles away, waiting to land. In a split second I had the feeling that the pilot of this plane had seen everything that I had. And I decided that what was best for him was to land immediately, in order to avoid [causing] him any more stress."

It took him several more minutes to clear all the traffic. Eventually, he says, "the situation was under control. There was nothing else to do. And then I just kept my calm for a few more seconds without hearing anything on the radio. All the pilots were respecting my sorrow. The job was done, so I just sat down on the carpet floor of the control tower up there, and I cried."

Logelin was in a unique position: he was both directly involved in the moment-by-moment crisis management after the crash and a witness to the disaster. Dozens of bystanders at the airport and rush-hour commuters on nearby roads also had witnessed various phases of the accident; many passengers, airport workers, and pilots had looked on in astonishment as the routine takeoff was transformed into a scene from a Hollywood disaster movie. Among them was American pilot Sid Hare, who later told a news reporter that he could hear Concorde's engines racing two or three times louder than usual prior to the sight of what he described as "a huge fireball, like a mini atomic bomb. One of the engines obviously had a catastrophic failure. It was trailing flames 200 or 300 feet." Other witnesses saw smoking debris fall from the plane as it lifted off. One airport employee, Julian Pyke, summed up the collective response of the stunned witnesses: "It was one of those experiences when everybody looks at everybody else with a 'It shouldn't be happening' sort of feeling."

The crippled Concorde had been clearly visible in profile on a trajectory that paralleled the north-to-southwest section of the four-lane N17 motorway, which lies west of Charles de Gaulle Airport and skirts the eastern edge of Gonesse and Le Bourget Airport. (The plane eventually crashed at the intersection of the N17 and the D902 roadways.) A grainy 15-second video of Concorde, taken by the wife of a Spanish truck driver using the same motorway, was later shared with news agencies. Her video tracks the plane as it judders along just above the treetops in the near distance. A huge plume of black smoke is seen in the wake of the burning plane.

Flight 4590 aflame: a plume of fire is seen extending from below Concorde's left wing in this still from a video shot from a truck on the N17 motorway, near Charles de Gaulle Airport, from which the plane took off. Moments later, the plane would crash in the town of Gonesse. (BUZZPICTURES / PREMIUM ARCHIVE COLLECTION / GETTY IMAGES)

"It passed 90 feet over us," another driver on the N17, Frederic Savery, 21, later told the *Telegraph*. "The whole back end was on fire. We saw it start to turn. All of a sudden, everything was black and we stopped immediately and called the firefighters." Another witness, 15-year-old Samir Hossein, had been playing tennis in Gonesse when Concorde fell from the sky. He recounted the last moments as the plane chopped off the tops of trees before rolling over and falling: "We saw flames shoot up 120 feet, and there was a huge boom."

From his post 9.5 kilometers (6 mi) distant, all that Logelin could see was that ominous column of black smoke. He could not tell where Flight 4590 had hit the ground. It had crashed into the small Hotelissimo, a hotel and restaurant in Gonesse, destroying the entire two-story wooden structure. The hotel had been nearly empty that afternoon, although a large group of British schoolchildren was due in later that same day. At the time of the crash, only one guest was present: Alice Brooking, a visiting Cambridge University student, who was employed as a guide by a tour company that catered to visiting students. She was in her first-floor room when she heard the sound of an approaching plane. "At first it just sounded like a normal takeoff," she recalls. "But then it got louder and louder and the floor began to shake. I said: 'My God, what's that?'"

Brooking opened her door, only to discover that the hallway was filled with flames. Quickly she closed it and retreated to the opposite side of the room. Through the window, a young man—a hotel receptionist—beckoned to her from the ground. She obeyed, jumping from her window and running barefoot across the grass to escape the inferno. Behind her, most of the hotel had been reduced to rubble. "There was just smoke and debris everywhere, and I inhaled some smoke. It was like being in an oven. I ran to some motorists and they shouted: 'It was Concorde, it was Concorde.'"

Not everyone in the hotel was so fortunate. Four women, members of the staff, ranging in age from 18 to 41, were killed instantly when the plane struck. Later they would be called the forgotten victims of the crash. All were chambermaids who had come to France in search of jobs to support themselves and their families. Ewa Lipinska, 18, and Paulina Cypko, 19, were both citizens of Poland, where they had lived in an industrial town near the Czech border. Both resided at the hotel and were relaxing in their rooms when the plane plowed into the Hotelissimo. Kenza Rachid, 19, was a dual citizen of Algeria and France. She had been hired the day before the crash, and it was her first day at work. The oldest of the four, Devranee Chundunsing, 41, was a native of the island nation of Mauritius and a single mother who would leave behind two orphaned children. When the plane struck, Rachid and Chundunsing had only 15 minutes left before the end of their workday: they were scheduled to leave at 5 p.m.

The Hotelissimo's manager, Michèle Fricheteau, 46, later told a reporter for a British daily that the other members of her staff of 12 men and women likely would have been killed as well had they not left the hotel around 3 p.m., wrapping up the workday earlier than usual after completing an early dinner setup for an incoming group of tourists, including the British schoolchildren. Fricheteau said that she had just finished dusting the reception desk alongside a trainee when she heard an "incredible noise." She speculated at first that one of the Concordes that routinely flew over the hotel was simply making more racket than usual. "I had barely finished saying this when I was hit by a huge ball of flames straight in my face. I immediately pushed my little trainee, saying to him to get out of the window. I turned around to go and get the others, and at that point everything collapsed in front of me."

•

In the days and weeks following the crash, Logelin struggled to come to grips with his role in the seemingly ordinary takeoff that had ended in catastrophe. "I would have given everything not to be there that day," he says, "but I was. I did my best for the crew, and the crew did its best to save the plane. Unfortunately, they did not make it." Although he was satisfied that he and the other controllers had fulfilled their duties, he was troubled by a lingering, ill-defined sense that he was, somehow, the lone survivor of the crash.

"I was hearing every day about the Concorde on television. And because I was the controller, I could not avoid being part of the story. So I was linked to this crash. I witnessed the crash, but I was also with them. Afterward there was just a feeling that I had of being the survivor of a catastrophe where there *were* no survivors. Which is quite a strange feeling."

3

A Lunar Landscape

The fiery crash of the iconic Concorde had unfolded with terrifying speed, and thus the precise second that the escalating disaster commenced was elusive. For pilots Christian Marty and Jean Marcot, it had begun when controller Gilles Logelin informed them, at 13 seconds after 4:43 p.m., that their aircraft was trailing flames: "Concorde zero, four-five-nine-zero. You have flames, you have flames behind you." Had some subtle event or mechanical malfunction preceded and possibly sparked the flames that Logelin had first seen beneath the wing on the left side and streaming behind the plane as it attempted to take off? Was that fiery tail—later described as being 61 meters (200 ft) long, or nearly the length of the fuselage of the plane itself—merely the first visible evidence of an accident already in progress? All these questions remained to be investigated. What was certain at this point, and yet hard to believe, was how quickly Concorde's attempt at flight had come to an end: exactly one minute and 18 seconds from the moment Marty began rotation, lifting the nose of the plane upward at 4:43:13 p.m., to the final moment of the crash, at 4:44:31 p.m. Paris time.

Firefighters and other rescue personnel were on the scene within minutes, prepared to provide emergency medical care to both those injured on the ground and any passengers who might have survived—although that seemed unlikely. Members of the media rushed to the crash site as well. Although kept at a distance, they were able to see and describe the burning pile of rubble that used to be a 45-room hotel and to send out reports about the makeshift field hospital that rescuers had set up nearby. It was soon apparent that there was little need for the field hospital; instead, the greater need was for space in nearby buildings that could be used as temporary morgues. Some victims' bodies were taken to rooms in the nearby Hôtel Les Relais Bleus in Gonesse, adjacent to the destroyed Hotelissimo. The Hôtel Les Relais Bleus remained intact and undamaged, although its owner and manager, Patrick Tesse, told the media that he "thought the plane was going to land on my desk. I saw it heading towards me in flames. I let go the telephone and ran." Most of the bodies, however, were taken to a makeshift morgue in a Gonesse auditorium, where plastic sheets were spread on the floor to receive them. All 100 passengers and nine crew members had been killed on impact. One guest at the Hotelissimo (Alice Brooking) and two members of its staff (manager Michèle Fricheteau and her trainee) had managed to escape, but the four chambermaids had perished, bringing the death toll to 113.

•

The crash brought forth an outpouring of grief and horror as well as heartfelt condolences from around the world. The names of those on the plane had not yet been released, since next of kin hadn't been notified. But word began to trickle in that most of the 100 passengers were from Germany, and that the flight crewmembers—the three-person cockpit crew and the six-person cabin crew—were all French nationals. The sheer size of the disaster and the fact that it involved one of just 13 supersonic passenger planes in service demanded an appropriate response from European leaders.

In Germany, which had lost the most citizens, all flags were ordered to be lowered to half-staff. "Germany is shaken and Germany is speechless," Chancellor Gerhard Schröder told 350 people who assembled for a memorial service in Hanover on Wednesday, July 26. As many of those in attendance wept, Bishop Josef Homeyer expressed their collective sadness and incredulity as he asked, "God, where were you in Paris?"

Other heads of state sent along messages of sympathy, including Queen Elizabeth and President Bill Clinton, who interrupted a news conference about failed peace talks in the Middle East to acknowledge the profound loss suffered by the people of Germany and France. British Prime Minister Tony Blair contacted French President Jacques Chirac and Germany's Schröder to express sympathy and acknowledge the terrible loss of life. In the aftermath of the crash, it would be reported that indeed almost all the victims, 96 of the 100 passengers on board, were Germans. One city in northwestern Germany had lost 13 of its residents, including six couples. One of the country's leading mass-market news outlets, *Bild*, ran a large photo of the crash scene with the stark headline "Ninety-Six Germans Burned to Death." Schröder declared that "Germany and France are united in their horror over the accident" and in their shared mourning for the victims. The four others on board, it would be revealed later, were an American Air France employee, an Austrian, and two Danes.

Schröder quickly dispatched his minister of transportation, Reinhard Klimmt, to the scene on the outskirts of Gonesse. Both French Prime Minister Lionel Jospin and Jean-Claude Gayssot, the French transport minister, immediately went to Gonesse to survey the scene, assess the damage, and consider what steps should be taken. "It's a terrible moment," said Gayssot. "There was nothing left to see there. It looked like a lunar landscape."

Reports were soon broadcast about people who had narrowly escaped the crash. Thirty-three German nationals who were booked to take the same cruise to Ecuador had been unable to obtain tickets on the doomed Concorde because of limited seating. Instead they had flown to New York on another Concorde before the crash. One well-known German actor, Günter Pfitzmann, then 76, had booked a ticket to fly on the chartered Concorde, but health problems led him to cancel at the last minute. "It is terrible to think that we might have been aboard that plane," his wife later told the BBC. "Our hearts go out to the families of the dead." Official calls for action were forthcoming as well; in addition to offering words of sympathy, politicians from all the affected nations were making it clear that they wanted answers about the cause of the crash. "We need to find out what went wrong, and soon," said British Conservative leader William Hague.

•

The age of the entire fleet of Air France Concordes and its prospects were uppermost in the minds of not only the investigators but also the aviation community, the public, and government officials in France, England, Germany, and the United States. In a CNN interview on the evening of the crash, Michael Goldfarb, former chief of staff of the Federal Aviation Administration, pointed out something that most in the aviation community already knew: "This plane is approaching its life limits," he said. He also revived a long-standing controversy about whether Concorde could generate profits that would justify its continued existence. "There hasn't been quite the market, as you know, for the supersonic that they originally envisioned. So it's questionable about the future life of the Concorde, whether they'll reinvest in it and keep it flying." Goldfarb nonetheless went on to praise Air France as "a very good aviation company" with a strong record "in their oversight of their industry." Concorde itself was "the icon of modern safe air travel" and "a world unto its own."

Immediately after the crash, Air France voluntarily grounded the remaining five Concordes in its fleet. Since only 20 Concordes had been built and only 13 of those were in service in 2000, the accident had reduced to 12 the number of Concordes operated in the world. The remaining five Air France Concordes were parked in hangars, where they would remain indefinitely until more answers were forthcoming. One Air France Concorde crewmember told London's *Observer* that flights should not resume until the cause of the crash had been determined. "I don't want to pilot Concorde again until I know what's happened," said pilot François Pradon.

But some officials, including Gayssot, insisted that every effort would be made to allow Air France to resume its Concorde flights. Others applauded the move by Air France to ground its Concorde fleet and called for British Airways to do the same—not for safety's sake but to put an end to sonic booms. "If the Concorde is finally to be grounded, the world may lose one of its icons, but at least it will be a quieter place as a result," wrote *Observer* columnist Richard Ingrams.

Air France, which had lost nine employees, offered to fly relatives of the German victims to Paris and assured the public that it would cooperate with investigators. It would pay for funerals and counseling for all who needed it, and it would award $20,000 to the relatives of each victim to cover immediate expenses. The airline also insisted that there were no undetected cracks in the wings of its Concorde fleet, an issue that was raised after the crash, since British Airways (BA) reported the presence of tiny cracks in the crossbeams of the wings of its Concorde fleet on Monday, July 24.

Across the English Channel, London-based Rolls-Royce's corporate stance was one of wait and see. It had little to say about what might have happened to Concorde's engines, other than that the company was in the process of gathering more information. British Airways reacted by cancelling its Tuesday-evening flights to and from New York, but on Wednesday, July 26, it resumed its regularly scheduled Concorde flights. After the crash, the engines on all BA Concordes were immediately inspected and found to be sound, said Mike Bannister, chief pilot of BA's Concorde fleet; as he told BBC News, "We looked right inside the engines to inspect them and make sure they were in tip-top condition." Concorde did not lack for outspoken defenders in the United Kingdom, including retired BA Concorde captain John Hutchinson. "It [Concorde] is in my view . . . the safest airplane that is flying in the skies today," he told the BBC. "For two reasons: one is that it's a very tough aeroplane, it's built in a very robust manner. And secondly it's got tremendous reserve capacity. By that I mean it's got a huge excess in capacity and power on the engines, all that sort of thing."

The French Bureau d'Enquêtes et d'Analyses pour la Sécurité de l'Aviation Civile (BEA, Bureau of Investigation and Analysis for Civil Aviation Safety) leaped into action, dispatching its investigators to the field and its bulletins, or *communiqués de presse*, to members of the media.

One of its first bulletins was issued on Thursday, July 27. The carefully worded press release refrained from drawing conclusions, instead laying out a simple, point-by-point factual overview of what had taken place. Much of what the bulletin reported was already known, but it did confirm that the plane had remained in level flight and was unable to gain altitude because of lack of thrust and the eventual failure of portside engines 1 and 2, combined with the drag created by the main landing gear legs and bogies, which had failed to retract. It also revealed that the plane's speed and altitude "remained constant" for the duration of the short flight—until the plane went into an uncontrolled roll just prior to the crash—and that the burning plane had left a trail of debris along the runway and "the entire length of the aircraft's path." Later it would be revealed that the plane had traveled just 9.5 kilometers (6 mi) from the threshold of runway 26R to the impact site in Gonesse.

In the same bulletin, the BEA pledged that representatives from all affected nations would be working with and for the French investigative agency. In addition, there would be two simultaneous ongoing probes, with different methods and goals. Like investigators for the American National Transportation Safety Board (NTSB), the French air crash detectives were called in to examine evidence and question people to determine the factors behind the accident without assigning blame. The other French investigation was judicial and legal in nature and overseen by three magistrates and a prosecutor, who would decide if any criminal charges should be filed in connection with the crash. A team of legal investigators would examine what had been done, or not done, by all parties involved, with an eye to determining if any laws had been broken and whether French air safety regulators had done their jobs; placing blame would be very much a part of their mission. They also were given authority to maintain control of the wreckage and the crash site, as well as evidence collected there and along the runway, including the flight data recorder and the cockpit voice recorder.

This approach was different from official investigations of air accidents conducted by government agencies in the United Kingdom and the United States. In those two countries, field representatives of the Air Accidents Investigation Branch (UK) and the NTSB (US) are the ones who work with police to collect and store evidence gathered at the scene. In France, however, the legal investigators are charged with collecting and storing evidence, therefore controlling who has access to it and when, setting the stage for a tug of war between the two groups with two different missions.

Chief among the questions being posed was why Concorde had not aborted its takeoff. Was it simply too late to do so? In other words, had Logelin's verbal alert that flames had appeared been received by the cockpit crew only after the plane had reached and passed V1, or decision speed, the speed at which a takeoff cannot be aborted? Had Captain Marty continued the takeoff to avoid colliding with other planes on the ground? Or because he thought he could use the airflow that was the by-product of the aircraft's speed to help extinguish the exterior flames? One thing was clear from the beginning: once in the air, Concorde was unable to gather the speed required to acquire lift and gain altitude; the usually high-flying plane had spent much of its brief time in the air just some 66 meters (200 ft) above the ground. Many witnesses said they saw the plane's wing dip to the left before it entered an uncontrolled roll and fell onto the Hotelissimo.

•

The day that Air France Flight 4590 was consumed by flames and fell from the sky, Yann Torres was on his motorcycle heading home from work, along with all the other commuters who shuttle around among Paris and its northern suburbs. Most live in the suburbs and commute south to jobs in Paris. But Torres lived south of Paris and worked in the northern suburbs at Le Bourget Airport, where the BEA has its main offices. Torres was an accident investigator for the BEA, an agency attached to the Ministère de l'Équipement, des Transports et du Logement (the Ministry of Equipment, Transportation, and Housing). As he headed south to Paris, he had a clear view of the sky over the accident site—a nondescript triangle of flat land occupied by two hotels and bordered by two wheat fields in their midsummer gold. Torres recalls:

"I saw all of this black smoke going up in the sky—very, very black." Its thick, noxious quality reminded him of the fumes released by burning rubber, which at first he guessed it was. "I thought that it was surprising to have a tire fire in that area right next to the runways. But when I got home, I learned that, in fact, it had been an accident with an aircraft." And not just any aircraft, but Concorde. "My first reaction was simply disbelief. The crash of the Concorde was something that no one would have imagined."

A seasoned accident investigator, Torres knew that no Air France Concorde, or any other Concorde, for that matter, had been involved in a fatal accident since the first models began carrying passengers in 1976. Like other Europeans, he considered it both a superior and a safe aircraft; its crews were widely regarded as the best of the best, and its operator, Air France, was respected as a leader in civil aviation around the world. Concorde's history suggested nothing that would lead him to suspect the catastrophe had been caused by something going wrong with the plane itself.

It would be Torres's responsibility to unravel the mystery of what had happened to Concorde's powerful Rolls-Royce/SNECMA Olympus 593 engines, which were designed for maximum efficiency at Mach 2. He was chosen to be the leader of the group tasked with studying structure, systems, and engines, one of seven working groups charged with finding and analyzing all evidence and information pertaining to the crash.

Another longtime BEA employee, Alain Bouillard, was similarly stunned by the news. "The day I learned that the Concorde had crashed, I felt two shocks," he says. "The first was the loss of this magnificent plane, this symbol of aeronautics—be it a French or British symbol—it was a symbol throughout the world. It was a magnificent plane." No sooner had he learned of the crash than he was notified that he had been chosen to lead the high-profile, demanding investigation conducted by the BEA. It was a huge responsibility, Bouillard recalled: "I then felt a second shock. It was when I was appointed as leader of the investigation. It seemed to me that it was an investigation that was extremely difficult, involving a plane that was very complicated. I remember it as if it were today—these two shocks—the loss of the plane and my designation as director of investigation."

But Bouillard had no time to ponder the implications of his appointment and the huge investigative task that lay ahead. He and other BEA investigators were at the crash site in Gonesse while firefighters were still hard at work laying down hose and trying to extinguish the inferno that had engulfed the two-story, L-shaped Hotelissimo and killed four women, whose bodies were buried in the smoking remains. "When we arrived at the location of the crash, the plane was still in flames. So we were not able to intervene," he says. "The priority was not our intervention. It was mostly the intervention of the rescue operations to see if there was anyone to save. The difficulty was the heat that reigned all around at the site and that prevented, on one hand, the intervention of rescue operations and on the other hand, our intervention to recover the flight recorders."

Foremost in his mind was the condition of the plane's recorders, including the cockpit voice recorder (CVR) and the flight data recorder (FDR). The FDR—the so-called black box—preserves the recent history of the flight by recording a stream of data collected several times per second, vital information about the performance of the plane's engines, controls, and systems. Flight data recorders are encased in metal boxes constructed to withstand the intense forces of impact and high temperatures. However, if the heat of the fire had been too intense, either or both devices might have been irreparably damaged, a development that would pose a serious setback to the investigation even before it had really begun. "We absolutely wanted to recover the flight recorders really fast," Bouillard says, "because they were in an area with very high temperatures. We were really worried about their condition and ability to give us the data for which we were waiting, because they could be deteriorated by the fire."

The crash site was a scene of devastation. Heaps of tangled debris, battered sections of the fuselage, detached tires, and the plane's massive engines were scattered across a large rectangular area 100 meters

The Gonesse crash site, overlain with the grid that investigators used as they scoured the area for evidence. The charred square in the left foreground is the remains of the Hotelissimo, where four employees were killed. (*BEA REPORT*, 69)

(328 ft) long from east to west, or about the length of a city block, and half as deep, 50 meters (162 ft) from north to south. The aircraft had fallen in a horizontal posture, practically flat and right side up, at a 120-degree heading and with "very little forward speed," according to the official BEA report. After impact, it broke apart, with most of the wreckage spreading generally to the south.

The crash site was located at the intersection of two busy roadways: the N17 and D902. Both were immediately blocked off. Bright-red tanker trucks, police cars, and other emergency vehicles were parked end to end on each roadway adjacent to the site. Immediately upon arrival, firefighters began training powerful jets of water on the burning hotel and wreckage, and a cloud of pale steam rose skyward. Several dozen gendarmes stood guard, ensuring that no trespassers could disturb the emergency workers or walk off with souvenirs that might in fact be important pieces of evidence. Later, after the flames were doused and the wreckage could be explored, workers wearing white jumpsuits and face masks arrived to spread tarps over the human remains, which silver-helmeted, red-and-black-uniformed firefighters then placed on stretchers and removed. By Wednesday, 80 bodies

Remains of Concorde's needle-shaped nose cone at the crash site in Gonesse. (ERIC BOUVET / GAMMA-RAPHO COLLECTION / GETTY IMAGES)

had been recovered, but the exhausting work of locating and removing corpses and body parts for identification would continue through Friday, July 28.

The force of the impact had created a shallow crater in which one of the 13 fuel tanks was buried. Not far away, to the east of what used to be the ground floor of the Hotelissimo, were parts of the landing gear and the outer, right-side section of the wing. Nearby, too, a single wheel lay embedded in the ground. Two more of Concorde's large wheels were found on the far western edge of the crash site. Part of the cockpit windshield rested on the far southern edge of the debris field, not far from the remains of the cockpit and Concorde's distinctive needle nose. Most of the rest of the plane lay in the center of the crash site, including the cabin floor and parts of the large delta wing. Nearly all of the fuselage had been badly damaged by fire, with the exception of the cockpit and nose cone, which were still bright white, reminders of what Concorde 203 had been in its pristine entirety.

•

As he stood on the edge of the debris field, Bouillard recalled a very different mental picture of Concorde, one he had formed during a trip he'd taken as a Concorde passenger. "I had that chance to fly as a passenger on the Concorde. I say 'that chance,' because aside from the extraordinary welcome by the cabin crew, I only retained this sensation of speed, the acceleration at takeoff, and the color of the sky at 60,000 feet. These are sensations that are very particular and unique that I haven't felt since and that I may never find again. I had this chance to live these very particular sensations that we can only get on the Concorde, and I will keep an indelible, and fantastic, memory of it."

Among the others who had both a personal and a professional stake in the crash was Jean-Louis Chatelain, then a pilot for Air France. Chatelain had been notified on the very day of the crash that he had been rated as qualified to fly Concorde as a captain. It was a major accomplishment, one that would advance his career and that validated the months of testing and training he had recently undergone before he could achieve the rank of captain. That afternoon, he recalls, "I was at home celebrating" and had not bothered to turn on the television to watch the news. Instead, he had been sipping a celebratory drink when he received a phone call from a friend. "Of course, many people among my friends knew I'd taken this Concorde training," he says. "I had a colleague, a friend of mine, who was based with me in the West Indies years ago. And he was the one who called me. He told me there had been a Concorde crash. So everybody knew in a few minutes that this happened—everybody—even the CEO of Air France," whose office window afforded him a view of runway 26R at Charles de Gaulle.

The news of the death of the captain and crew, all well known to Chatelain, was difficult to absorb. Marty, Marcot, and Jardinaud were fine men and respected coworkers in the Air France family, and Chatelain and Marty were more than acquaintances; they were friends. "You can easily imagine that it was a total shock because I knew everybody," recalls Chatelain. "How can I say more than this? We were all totally shocked." The story of Marty's and Marcot's heroic efforts to save the doomed plane spread quickly as well. Chatelain and Marty, like all professional pilots, understood the risks of flying but trusted their airplanes, coworkers, and shared expertise to keep them aloft and alive. Yet the reality is that even the best pilots cannot master every situation or reverse every unfolding calamity. "Like many other pilots, I've tried to put myself in their situation," says Chatelain. "At one point, they knew they were going to die, but there was no panic at all. To the very end, they tried to find a solution."

Chatelain would soon attend a private gathering called by Air France executives at the corporate office at Charles de Gaulle. "There was really a big internal meeting with all of the Air France staff who worked at the airport," during which Jean-Cyril Spinetta, the airline's CEO, did his best to reassure his stunned employees. Air France would pay for funerals and provide for the immediate needs of the families of the Air France victims, he said. The tearful gathering was followed, later that week, by a formal memorial service at l'Église de la Madeleine in Paris's Eighth Arrondissement. In attendance were French government ministers and an official delegation from Germany led by Foreign Minister Joschka Fischer, along with uniformed Air France pilots and flight attendants who acted as honor guards. They stood to either side of the stone steps of the 19th-century Roman Catholic church in a show of respect as the victims' relatives, clutching one another for support, climbed the stairs and passed between the church's Corinthian columns. A candle for each of the 113 victims was set on the altar, and the moment of silence that marked the beginning of the service was simultaneously observed at the three airports in Paris. "It was totally emotional," says Chatelain, who attended the memorial service at the historic church. "But you cannot sustain this level of emotion. This is life, you have to look forward."

For Chatelain, the near future included a demanding assignment that would require him to study everything that had taken place in Flight 4590's cockpit on the afternoon of July 25 and to consider which decisions were made before and during the minute and 18 seconds that elapsed between 4:43:13 p.m., when air traffic controller Logelin radioed the report of flames, and 4:44:31 p.m., the end of the CVR. Eight years earlier, Chatelain had been called upon to help investigate another major French aviation accident: in January 1992, an Airbus A320 flying at night from Lyon to Strasbourg had slammed into a mountain. Eighty-seven people onboard Air Inter Flight 148 were killed, including the pilots; a flight attendant and eight passengers survived. One of the key questions in that crash had been whether the pilots had correctly programmed the plane's computer and how that had affected the aircraft's rate of descent. Chatelain, an expert in cockpit flight operations, was appointed by the French minister of transportation to analyze the pilots' performance and training to operate the plane's guidance system. At that time, he recalls, he was "the first 100 percent airline pilot to be appointed on such a commission, and the experience was positive." He spent the next year working with the BEA and became part of an ongoing accident investigation committee. In the wake of the Concorde crash, he says, "It was almost natural that the minister of transportation appointed me again."

During the investigation, Chatelain would have cause to think back to an upbeat conversation he had with his friend Christian Marty many months before, not long after Chatelain was informed he was eligible to begin the rigorous months-long training program that he had to complete before being certified as a Concorde captain. As part of his training, Chatelain knew he would have to spend many hours in a Concorde flight simulator and would also be required to execute actual supervised takeoffs and landings. His handling of the aircraft during every phase of takeoff would be watched and evaluated. Chatelain was excited to begin formal training, and a bit anxious as well. "I called him [Marty] and I asked him, 'What's the trick? What would you say after getting your Concorde rating?' He told me there is one thing to remember: 'You need speed.'" That simple but telling remark would remain with Chatelain in the months to come.

•

Several thousand miles away, in Brazil, NTSB investigator Bob MacIntosh was on assignment when he heard the news out of Paris. "Even the local news media down there was alive with the sad news that the aircraft had crashed. And I recognized that it was probably going to be a very important day in aviation history. When a Concorde faltered and fell to the ground, especially with the video that was available of an aircraft on fire, it shook the world. It really did."

Although the plane was operated by a French airline and the crash took place in France, there was an American connection that called for involvement by the NTSB. The tires on all Air France Concordes had been manufactured by Goodyear in the United States. (British Airways Concordes were equipped with tires made by Dunlop Aircraft Tire Co.) There had been problems with the tires on Concorde in the past. A series of serious incidents involving burst tires, holes in fuel tanks, and debris from blowouts ingested into engines had plagued Concorde from 1979 through the early and mid-1980s. In one incident in June 1979, two tires had blown out on a Paris-bound Air France Concorde taking off from Washington Dulles International Airport, near Washington, D.C. Exploding tire and wheel fragments damaged one engine, punctured three fuel tanks, and tore a hole in one wing before Captain Jean Doublaits safely landed the wounded jet. "This 1979 event was a real wakeup call," says MacIntosh.

Similar events followed at Dulles and John F. Kennedy International Airport over the next two years. As a result, the NTSB got involved. It conducted a study and sent a letter to the BEA voicing its concerns and warning of the dangers posed by tire blowouts, including fires and explosions that could result from a fuel leak. "As a result of this study that looked at the 1979 to 1981 events with tires, there were five or six actions that were undertaken, including inspecting the tires very carefully before each takeoff, looking at the tire inflation pressures, looking at the automatic braking system and doing things to improve the integrity of the tires themselves," says MacIntosh.

In 2000, it made sense for a representative of the NTSB to take part in the investigation in Paris, which would be overseen by the BEA. Still, MacIntosh was very much aware that his involvement was another sign of both the scope of the investigation and its unusual nature. "It was unique that a French airplane [investigation], on French soil, would have a United States representative," says MacIntosh. "But the United States manufactured and designed the tires, and consequently we in the United States were tasked with being an accredited representative to the official investigation. My job was to get together with the manufacturer, Goodyear, and with the FAA and establish a team presence in Paris." MacIntosh was quickly notified of his new assignment, and he immediately left Brazil for France and the city of Gonesse.

"When I arrived at the crash site, it was still in chaos," recalls MacIntosh. But at least the entire area had been secured and preserved, which was important to the investigators who would be charged with forensically reconstructing the crash. The local gendarme had been in charge from the beginning, "so things really hadn't been moved that much. And we could recognize the larger sections of the wreckage." Almost everything was charred and dripping wet with the tons of water used to douse the flames. It was a forlorn spectacle. Pools of sooty water had collected in low spots. Clothing and other miscellaneous items from dozens of bags and suitcases were strewn everywhere, lying in heaps and mingled with rivets and bits of plastic. A white bathtub from the Hotelissimo sat awkwardly atop a pile of burned lumber.

The area had been cordoned off with red-and-white-striped accident scene tape, and traffic cones were used to mark the places where bodies were found. Removing the human remains was a painstaking and depressing undertaking. The crash had been indiscriminately destructive, shattering the plane and leaving few bodies intact. "Once you got used to looking through the smoke, the horror hits you," an emergency worker told Agence France-Presse. "By the dozens, bodies or pieces of bodies, horribly mutilated. Here and there shoes, books, parts of suitcases that are wasted away."

What struck MacIntosh, who studied the crash site with a professional's critical gaze, was the degree to which the force of the impact had scattered various parts of the plane to the edges of the debris field. "The wreckage site was what we call a 'typical *plus*,' in that sometimes we find a very concentrated area. This time we didn't," he says. "Things had bounced and found their way into more distant spots. There was a lot of it spread out, and we also had a building involved that had been set on fire. We had very large tires and landing gear—parts that were spread around." In addition, the debris "had been partially rearranged" to allow retrieval of the victims' bodies. "So there were some tremendous challenges" when it came to reconstruction. Like Torres and Bouillard, MacIntosh knew that the plane's cockpit voice recorder and flight data recorder were vital "because they would give use some good indications of aircraft performance. But we were also looking for any physical clues that we could find in that wreckage."

The investigation would be conducted primarily at two very different locations: large, well-ordered Charles de Gaulle Airport, where the plane had taken off, and the small, deeply scarred crash site. Different teams of investigators would focus on one site or the other, depending upon their areas of expertise, and they traveled back and forth between the two. "When we were forming investigational teams," says Torres, "there was one team that was in charge of describing and detailing the accident site. I wasn't part of it, so I only went to the crash site on the second day. And indeed, on the second day we realized that we had on our hands a plane that was totally destroyed by the flames, by the fire, and whose parts were mixed with those of the building it had crashed into."

Because the accident had unfolded in broad daylight during rush hour in one of the most populous metropolitan areas in the world, the investigators would be supplied with a plethora of raw evidence and information provided by the public. Dozens of witnesses had seen the plane catch fire at the airport. Some said they saw parts fall away before the plane lifted off and reported that they had noticed debris on the runway. These latter statements were easily verified. Several witnesses said they had seen it bank sharply to the left not long after leaving the ground. Differing accounts were given of the plane's movements in its final throes. One witness, a lorry driver, said that just before the plane fell to earth, it had flipped over, exploded, and split in two. Others described watching the plane bank to the left and roll over before striking the hotel nose-first.

Photos of the aircraft in various stages of distress had been taken from multiple locations and points of view. "The videos and the photos that were taken were really important to us," says Bouillard, "because they were elements that permitted us to estimate the fuel's rate of flow under the wing of Concorde. These equally allowed us to determine where the fuel leak was situated. So the videos and the photos that were taken by different witnesses were one of the important elements for the investigation."

Among the witnesses to the series of events leading up to the crash was Jean-Cyril Spinetta, the Air France CEO, who told reporters that he had seen Flight 4590 take off as he looked out his large office window on the third floor of the airline's headquarters at Charles de Gaulle. He too verified that flames were streaming from behind and beneath one or more of Concorde's port engines as it left the runway.

Bystanders' photos included those taken by Andras Kisgergely, a 20-year-old student and amateur photographer from Hungary. He and a friend had been on a plane-spotting trip around Paris, and Kisgergely had been sitting in his car near Charles de Gaulle, waiting for his friend to finish photographing another plane parked at a hangar. Kisgergely was looking out his open driver's-side window when the stricken Concorde came into view shortly after takeoff, flying over the adjacent runway. At the same time, he heard a noise that he described as sounding like "burning gas from a tank," He grabbed his camera and began snapping photos, which he later sold to Reuters. The shots were published in newspapers around the world. In them, Concorde can be seen flying parallel to the runway, slightly higher than the trees in the foreground. Concorde itself is a ghostly gray, but the flames streaking behind it are bright yellow and as long as the plane itself.

As noted earlier, at least one amateur video, taken on a camera phone by the wife of a truck driver on the N17 highway, surfaced as well. (The couple later released the video but asked their names not be made public.) The grainy but telling clip shows Flight 4590 moments before it crashed, flying along 61 meters (200 ft) off the ground on a trajectory roughly parallel to the motorway. The Concorde is on fire and leaves in its wake a long trail of black smoke (see page 27). Others had watched as the burning aircraft entered an uncontrolled roll and fell on top of the small hotel, which sat between two wheat fields, sending up a plume of smoke that could be seen as far as 16 kilometers (10 mi) away. Many wondered if the pilot had been aiming for a stretch of flat, empty farmland but fell short of his hoped-for landing zone.

It wasn't just the media whose nonstop reporting kept the Concorde story on the front pages of the world. On Thursday, the residents of Gonesse took to the streets in a public display of mourning. Several hundred citizens, joined by the mayor and local dignitaries, locked arms and marched in a silent procession from city hall to the accident site to set down bouquets of white roses in memory of the 113 victims. Among the mourners was the manager of the Hotelissimo, Michèle Fricheteau, her burned arms wrapped in bandages. News photos showed her holding a single white rose as she stood near the ruins of the hotel, her adult children supporting her.

For Gilles Logelin, such photos on the front page of Paris newspapers and similar reports on television and radio were profoundly unsettling but also impossible to ignore. "If this had been another plane, it might have been different," he says. "But this was the Concorde. So this crash was everywhere in the media. It's not easy when you're the air traffic controller because the next day, or even a few days after that, even if you don't want to watch the news on the television, you have the temptation to do so. And then you hear that the controller said this or did that. It's a difficult situation to deal with, because the other controllers and I were not used to that."

The pressure was on to solve the mystery of what had happened to the iconic Concorde. All four men—Torres, Bouillard, and Chatelain, working for the French BEA, and MacIntosh, representing the NTSB—were well aware that the stakes were high and the world would be watching. "Because we were dealing with Concorde, there was media pressure present, since it was a plane that was very well known," says

Torres. "It was the only supersonic plane, and this was the first accident with fatalities for the Concorde. We knew that we were facing a very, very serious event."

Concorde was not simply a unique plane but also a symbol of French and British achievement and excellence. "It was a plane that everyone knew and that was, how to say, an emblem of French aeronautics—something very particular and of which all the French, and I think the British, too, were proud," says Bouillard. As the lone American investigator, MacIntosh knew it was vital to all the parties involved to consider all the factors behind the accident, whether tangible or intangible. "The investigation brought into focus, very early on, national pride for all involved—for Aérospatiale, the manufacturer; the French flag carrier, Air France; and the pride of the British, who had a great deal to do with the design and had furnished the Rolls-Royce power plants for the aircraft," says MacIntosh. "So there was national pride on both sides of the Channel that was involved in this accident."

•

In the wake of the Concorde crash, all of France was caught up in both a state of mourning and a sense that the crash marked a decidedly dark day in the annals of aviation. An entire generation had come of age along with Concorde. "Its image is everywhere in France," observes Logelin. "When you are at school, you learn about Concorde. As an adult, you see it on television, and it makes you dream about going faster, going further. So it's something that is in the collective imagination." Many in France feared that the reputation of the *l'oiseau blanc*, the white bird, a symbol of French excellence, would be permanently tarnished. And the crash of Concorde would be the death knell of the time when air travel had opened up a world of glamour and unlimited possibilities. But others retained their faith that Concorde's reputation would survive the crash, and that it would be remembered and celebrated as one of the most important planes of the 20th century.

From the day of the crash onward, the investigation would be colored by a conceptual polarity that existed between two disparate and conflicting versions of Concorde itself: the high-performance supersonic aircraft that was an exquisitely complex and brilliant piece of engineering, one of the great technical achievements of modern commercial aviation; and the spectacle of the malfunctioning jetliner that had erupted into flames for no apparent reason upon takeoff and lumbered along for a little more than a minute before plummeting to the ground, leaving behind a jumbled puzzle of melted metal and charred fragments—the reconstruction of which presented a forensic nightmare for investigators. "What made this investigation complex, first of all, was that the plane was extremely complicated technically, and with very particular flight qualities," observes Bouillard. "That which made the investigation even more complicated was this incomprehension of the fact that everyone had seen the plane fly with enormous flames toward the rear. And that it was destroyed by fire, so that there were very few [undamaged] pieces left. And all of that appeared as something really complicated."

Investigators would be challenged to look beyond their considerable collective knowledge of subsonic planes—their working parts, design, and maintenance and how they flew—and to school themselves in the very different structural and operational characteristics of Concorde. "There were a lot of things that made the investigation of the Concorde challenging," says MacIntosh. "Its performance, and the way that it used much higher speeds [than ordinary jetliners] to lift off and to take off, its higher angle of attack as it climbed out. Those were unique." In addition, "many of the materials that are used on the Concorde were manufactured specifically for that aircraft. Consequently, those materials were very special and needed to be looked at in a special way. Even the way the crew was trained was unique. So there were a lot of things that made the investigation much more complex than ones that we had done previously."

Only a few could claim mastery of the extensive and arcane body of engineering and technical knowledge that made Concorde fly and kept it in the air. "It had a very particular style and silhouette which made it extremely rapid, since it was the only commercial transport plane that could fly more than two times the speed of sound," says Bouillard. "It was also the forerunner in numerous categories, such as the electric commands." The investigators would use a divide-and-conquer strategy to analyze its systems, working parts, and technical characteristics from end to end.

As usual, the investigators would begin by asking questions. A fuel leak seemed likely, but what had caused it? Was there any connection between the fatal crash and Concorde's history of tire blowouts? What about the last-minute replacement of the defective engine part that Captain Marty insisted be performed before takeoff, delaying the flight? Had Concorde been brought down by something as prosaic as human error or a mundane mechanical failure tied to maintenance of its engines? Or was there a deeper flaw in its elaborate design, something that might put other passengers on other Concordes at risk?

•

As the investigators began collecting evidence and interviewing witnesses, speculation abounded about Concorde's viability, the future of supersonic air travel, and whether the crash, combined with the realities of 21st-century commercial air travel, had dimmed or irreparably diminished the original vision of Concorde: a supersonic plane that could connect the people and cities of the world in a way never before realized.

Perhaps humankind itself should take heed, as one opinion-piece headline in the *Guardian* suggested: "The Lesson of Concorde Is That We Can't Go Any Faster." The writer, Martin Woollacott, ruminates on the historical and cultural forces that gave rise to Concorde and the 20th-century vision of mechanized progress, which centered on the idea of speed as a means to overcome the barriers that impede humankind's progress toward a seemingly limitless future. "Advances in motive power were for a long while the main way in which progress and national competition in technology were measured," he writes. "First at sea, then on the railways, then on the roads, in the air and finally in space, more and more rapid movement was seen as an unalloyed good and also, in some hazy way, as a key to a fuller understanding of the world. All kinds of speed were interconnected, forming a transmission belt to a future which people believed would be radically changed." Yet the crash of Concorde, the fastest passenger plane in the world, would once again revive controversies that had dogged the plane since its inception and which had never entirely been put to rest.

By the early 1960s, when Britain and France signed a historic pact to develop a supersonic jet, the race was on, and every schoolchild was brought up believing in the inherent virtue and the thrill of going faster for its own sake. "The front of the *Boys Own* annual of half a century ago would typically feature a speeding train in the middle ground, a fast airplane above and a racing car in the foreground," writes Woollacott. "Such icons are still with us." Flying faster became both a goal and an Icarus-like compulsion that drove governments as well as individuals. "Disentangling the genuine advantages of speed from its cult aspects has always been a problem," Woollacott opines. "This was certainly the case in the era in which the Concorde was conceived." He ends on a note of caution: "Man can go faster, but that does not mean it is necessarily worth doing so."

4

From Dream to Reality

The paper airplanes that were the first models for Concorde were held together with sticky tape and lofty intentions. Fashioned by mature men with youthful dreams, the handmade planes—or paper darts, as the British called them—were meant to be tested in a wind tunnel at the Royal Aircraft Establishment (RAE) in Farnborough, Hampshire. Still, many of the models made impromptu maiden flights over the English countryside on pleasant, wind-free days. "There's no better way to test an idea than to take it outside and see if it flies," says 77-year-old Alan Perry, one of the group of Concorde engineers and scientists who worked for Bristol Aeroplane Company in Filton, near Bristol in southwest England, in 1959, where the earliest conceptual work on supersonic jets was then beginning. He and his colleagues would "fold them up [the paper planes], take them outside at lunchtime if the weather was nice, and see who could fly them furthest from the hangar," he recalls. "Sometimes we'd even use our punch cards."

The models were created to try out various wing designs for a supersonic jet. The most promising ones were suspended in the RAE's 7.3-meter- (24-ft-) long wind tunnel and subjected to high-speed blasts of air to see how they performed, explains Peter Turvey, former senior curator at the Science Museum at Wroughton. Turvey exhibited the collection of yellowing papier-mâché planes in 2007 after they had been discovered in a dusty warehouse, where they had languished for more than four decades. Someone had thought to keep some of the models, created under the direction of English scientist W. E. Gray, who led the group of designers that included Perry. Not much more impressive in appearance than the product of a child's fancy, the models were nonetheless "a crucial stage in the plane's development," says Turvey. "This was real boffin stuff. It was a case of shouting, 'I have an idea!' and then giving it a go."

•

The seeming insouciance of the British designers bordered on audacity, and it enabled them to embark upon a fantastically difficult project whose chances of success seemed slight despite the combined engineering talent of two nations, Britain and France, and their shared commitment to design and build a supersonic passenger plane. "If there is one word that best describes the Concorde, it is *advanced*," says retired Concorde pilot Jean-Louis Chatelain. "At the time it was designed—in the sixties—it was almost unbelievable that they dared to address such a challenging design. If you look at current aircraft technology, such as the Airbus technology, you can see it borrowed a great deal from the Concorde, which was a kind of laboratory for aircraft manufacturing in general."

Over the course of 10 years, the British Aircraft Corporation (BAC) and Sud Aviation (later Aérospatiale) would jointly develop and manufacture a revolutionary passenger plane capable of flying safely and smoothly at more than twice the speed of sound. Both an engineering marvel and a wonder to behold, the first Concorde prototypes took to the skies in 1969. Anyone familiar with the story of Concorde's inception and development, however, knew the counterintuitive secret to what would prove to be its age-defying beauty: Concorde was never intended to make a statement about how a supersonic aircraft *ought* to look. Instead its futuristic form was the incidental consequence of the demanding aerodynamic and mechanical requirements of supersonic flight and the inspired solutions made possible by the insights of a mathematical genius, Johanna Weber, and the dozens of English and French engineers who put her ideas into practice. Every structural and working feature of Concorde, from its peculiar droop nose to its pointed tail, owed its existence to solutions devised in answer to the rigors of supersonic flight, with concessions made to the related but very different problem of how to make Concorde behave like other planes—that is, able to take off and land at *sub*sonic speeds.

It was the first commercial airplane with carbon brakes and a special system created to move fuel between multiple tanks, achieving pitch control by shifting the plane's center of gravity. And it was the only plane in which it was possible for a passenger to look up from a meal of lobster thermidor served on a linen tablecloth and marvel at the sight of a fighter jet cruising alongside at supersonic speeds. Concorde was, in many ways, an homage to what was then perceived as the future of flight. But it owed its existence to the early, perilous exploits of pioneers in the field of aviation. None of the aerodynamic innovations built into Concorde could have evolved without the advances achieved at great cost during the heat of World War II and its aftermath, when a generation of daring test pilots put their lives on the line to prove that the experimental planes they were flying could go faster than any aircraft had ever gone before, fast enough to exceed the speed of sound.

Even at its outset, the 20th century had announced itself as an age defined by acceleration. All that was slow and incapable of going faster seemed destined to be left behind in the race to keep up with fast-moving events, transformative inventions, and new motorized conveyances on land and in the air. The emerging field of aviation and the science of aerodynamics contributed immeasurably to a growing infatuation with speed that promised to change the way humankind traveled, as well as how people experienced distance and time. Physicists and engineers would be the first innovators, followed by the pilots who flew experimental aircraft and set new airspeed records.

In the 1920s, the Swiss-born aeronautical engineer Jakob Ackeret helped Germany lead the way with university-based research into wind tunnels and the effects of supersonic airflow over wings. Near the end of the decade, the first rocket-powered plane—the Ente, meaning "duck" in German—flew 1.2 kilometers (0.75 mi) after being launched from a popular gliding spot in the Röhn Mountains in Germany on June 11, 1928. The plane's owner was rocket enthusiast Fritz von Opel, grandson of Adam Opel, founder of the Opel automobile company. Undeterred by the Ente's crash landing, von Opel commissioned a second rocket-powered plane, the RAK 3, a glider powered by 16 rockets, and piloted it himself before a crowd of spectators in Frankfurt-am-Main on September 30, 1929. He flew 1.5 kilometers (0.93 mi) before making a hard but successful landing that demonstrated that rockets could be used to power a fixed-wing aircraft.

In 1935, leading European scientists gathered for the fifth Volta Conference in Italy, a congress devoted to *le alte velocità in aviazione*, or high-speed flight in aviation. By that time, Italy's national air force had created the world's only high-speed flight research squadron. Conference attendees took a field trip to an aerodynamics research center at Guidonia, near Rome, and toured laboratories where scientists were collecting data on supersonic flight.

Even before the start of World War II, far-thinking scientists, notably Adolf Büsemann of Germany, were pondering the possibility of jet engines and supersonic flight as well as the advantages of swept-back wings over straight wings for flying at high speeds. The Germans became the first to employ the combination of swept wings and jet engines in military aircraft when they introduced the rocket-powered Messerschmitt Me 163 Komet and later, near the end of the war, the Messerschmitt Me 262 fighter, powered by an early version of the centrifugal-flow turbojet engine design invented by English pilot and engineer Frank Whittle. During World War II and shortly thereafter, powerful propeller-driven fighter planes, such as the British Spitfire and the German Messerschmitt Me 209, while making high-speed dives, began to approach the speed of sound, or Mach 1—about 1,169 kilometers per hour (727 mph). (The actual speed of sound is not fixed but varies according to conditions, including altitude and air pressure.)

However, before they could reach Mach 1, pilots had to pass through the previously unexplored speed-regime range that straddles the speed of sound. Despite its name, the "sound barrier" is not a wall-like obstacle but rather a gulf that must be traversed between the subsonic and supersonic regimes; it might more accurately be called the shock zone. This zone, which begins at speeds of about Mach 0.8 and extends to about Mach 1.2, became known as the transonic range. The shock waves that pilots encountered in this speed-regime range were unknown at the time, because there were no wind tunnels available that were capable of testing models at speeds sufficient to reveal the shock waves. Two issues confronted aircraft as they approached the speed of sound: they needed increased engine power to overcome increased drag, but they also needed new aerodynamic shapes to overcome the loss of lift on their planes' wings and control surfaces due to the shock waves that develop at transonic speeds. Increased engine power would prove to be the easiest problem to fix. Overcoming shock wave problems was far more difficult.

The transonic speed range was initially identified only because of the strange things that happened to aircraft when they entered it. Suddenly and inexplicably, aircraft controls became unresponsive to the pilot's commands. All that pilots knew was that as their planes approached the speed of sound, frightening and even potentially fatal control problems caused their normally responsive aircraft to behave unpredictably. Sometimes a plane would nose over and spin to earth in an unrecoverable dive that came to be referred to as a compressibility stall, or simply Mach tuck: the aerodynamically induced downward pitch of a plane's nose as airflow over the wings exceeds the speed of sound before the plane itself has reached supersonic speeds.

Unbeknownst to early test pilots, at speeds around Mach 0.8, invisible shock waves formed along the top surface of their aircraft's wings. As these shock waves moved across the wings, from the leading edge to the trailing edge, they shifted the center of lift of the wing rearward. This change in lift position caused the aircraft to pitch forward, nose-down. Aircraft without horizontal stabilizers and elevators strong enough to resist would then enter an unrecoverable dive. The term "Mach tuck" itself was inspired by a popular maneuver in the sport of competitive diving that is known as a forward tuck: a diver performing this move folds herself into a ball and hugs her knees while spinning into a forward somersault. After several rotations, the diver straightens out at the last second and slides smoothly into the water. But planes that underwent similar gyrations were headed for the unforgiving ground and destruction. It would take years of testing and experimentation with radical new airframe and wing designs to pave the way for supersonic aircraft such as Concorde.

•

A new breed of aviator, the test pilot, would emerge to fly these experimental aircraft, which were built to break records and inaugurate a daring new idea: a plane that could fly faster than the speed of sound. *Test* was the operative word in these years, as both pilots and planes sometimes were lost to the mysterious forces generated by shock waves. In November 1941, Lockheed test pilot Ralph Virden flew his twin-engine P-38 Lightning through a dive test and fell victim to the Mach tuck effect—when his plane hit Mach 0.675, the airflow over its wings became supersonic. Virden was killed in the ensuing crash, during which the plane's tail was ripped away from the fuselage.

Some planes and pilots were able to survive the Mach tuck effect, enabling them to return with detailed accounts of their experiences. In 1943, Tony Martindale, a test pilot for Rolls-Royce, was instructed to fly the British-made Spitfire PR Mark XI to a height of 12,192 meters (40,000 ft) above sea level, where he would initiate controlled—but incredibly risky—power dives. During these dives, the whole aircraft would "buffet and shake, sometimes to the point of structural failure," explains Concorde pilot and author Christopher Orlebar. Martindale also experienced "strange effects when diving at such speeds" and observed a dangerous tendency for the Spitfire's controls "to do the reverse of what was expected of them." The reason? Unseen shock waves in the transonic range were discombobulating the control systems of his plane by disrupting the flow of air across the wings and control surfaces.

Lessons learned by test pilots and advances in design features of military aircraft that had been forged in the heat of war were later put to use in the civilian sector, with mixed results. On September 28, 1946, British newspapers ran dramatic stories of the death the previous day of the chief test pilot for the De Havilland Aircraft Company, Geoffrey de Havilland Jr., son of renowned aircraft designer Sir Geoffrey de Havilland. De Havilland, 36, was killed when the De Havilland DH 108 he was flying exploded and crashed into the Thames estuary.

At the time, De Havilland was actively preparing himself and the plane for an official attempt at breaking the world speed record. A search was made for his body, which was found; he had died as a result of a broken neck attributed to the violent shaking of the plane during a high-speed dive. News accounts reported that de Havilland was said to have "attained true level speeds considerably in excess of the world record speed of 616 miles per hour [991 km/h]" while piloting the experimental aircraft, known as the Swallow for its distinctive 43-degree swept wings.

One of the first pilots to actually detect the invisible killer shock waves was NACA experimental test pilot George E. Cooper, who was conducting test flights in a P-51 Mustang when he noticed sunlight striking a wing at an unusual angle, exposing a ghostly shock wave creeping across the wing—an effect made possible by the refraction of sunlight through the changing density of the airflow. The fascinated Cooper, who later took photographs of the same phenomenon, noticed that he was able to make the shock wave shift back and forth, from the leading to the trailing edge of the wing, by increasing and decreasing his speed. As the wave moved, an accompanying buzz from his instruments warned that he was encountering instability in the air flowing over the control surfaces of the P-51's wings. Cooper had made the vital connection between the mysterious shock waves and the loss of control that plagued aircraft when they entered the shock zone.

•

Like other test pilots, Cooper was achieving high speeds during controlled dives. That risky tactic would become obsolete, however, as propellers, such as those on the P-51, were replaced by jet engines. The jet engine would revolutionize both military aircraft and civilian passenger planes; it would also make manifest the dream of supersonic flight. At the conclusion of World War II, the victorious Allies had studied German engine designs and launched the jet age. By the 1950s, nearly all military combat planes were powered by axial-flow turbojets. By the 1960s, civilian aircraft followed suit, with large commercial aircraft makers, including Boeing and McDonnell Douglas, abandoning piston-powered, propeller-driven designs for turbojet designs.

It quickly became apparent that jet engines had the power to boost aircraft from subsonic to supersonic flight speeds. However, crossing through the transonic speed range into supersonic flight required not just power but also entirely new airframe designs. The physics of flight in the supersonic regime were radically different than the physics governing subsonic flight. New wing designs were required; wings needed to be thinner and swept back to reduce the drag encountered at higher speeds.

But in October 1947, no jet engine yet built was powerful enough to fly faster than the speed of sound. Instead, an experimental rocket-powered engine was used to enable American test pilot Charles (Chuck) Yeager, piloting the Bell X-1 on October 14, 1947, to become the first pilot to surpass the speed of sound by flying at Mach 1.06, or 1,127 kilometers per hour (700 mph), at an altitude of 13,000 meters (43,000 ft) above Rogers Dry Lake in California's Mojave Desert. Yeager was elated but also surprised by the unexpected, "smooth as a baby's bottom" experience of supersonic flight. "Grandma could be up there sipping lemonade," the famously laconic Yeager later said of the easy ride that had awaited him on the other side of the so-called sound barrier. (Yeager enjoyed a still more burnished reputation once it was revealed he'd made his legendary flight with two cracked ribs, the result of an impromptu moonlight horse ride a couple nights earlier.)

The Bell X-1, which Yeager named *Glamorous Glennis* as a tribute to his wife, was more durable than beautiful. It had short, stubby wings that fit inside the bow shock wave, the curved, drag-producing wave that forms as a plane's nose plows forward at supersonic speeds. *Glennis* also featured elevated horizontal stabilizers at its aft end, which set them above the shock waves generated by the wings.

On August 20, 1955, American Colonel Horace Haynes, a combat pilot and veteran of World War II, took off from Edwards Air Force Base, in Southern California, in an F-100 Super Sabre 53-1709. The plane reached Mach 1.24 at 12,192 meters (40,000 ft), setting a new international speed record of 1,323 kilometers per hour (822 mph). This was both the first supersonic world speed record and the first speed record set at high altitude. By the 1950s, the British had begun incorporating the shock wave–resistant delta wing into their own supersonic research aircraft, including the Fairey Delta 2, or FD2. In 1956, British test pilot Peter Twiss broke Haynes's short-lived world speed record by flying an FD2 at more than 1,609 kilometers per hour (1,000 mph) in level flight. Between them, the two pilots had done away with the previously dominant notion that aviators and their planes were incapable of surpassing the speed of sound and surviving.

The production of supersonic military aircraft that followed and improved upon the rocket-powered Bell X-1, including the F-100 Super Sabre and the FD2, involved turbojets almost exclusively. Turbojet engines were soon shown to be capable of developing similar thrust, but they used less volatile fuels, such as kerosene. Also, instead of separate onboard liquid oxygen tanks, the turbojets used oxygen from atmospheric air as an oxidizing agent. The improving turbojet engines were simpler and safer and enabled longer-range flights with superior fuel efficiency. Even before advances in jet engines had made the idea of supersonic flight feasible, some revolutionary thinkers in Britain were actively studying the possibility of building a supersonic plane that could carry ordinary travelers from London to New York.

•

In 1956, the same year that Twiss set a new world supersonic speed record, a group of British proponents of civil supersonic transport did something almost as bold but not as well publicized. They formed a working committee to investigate the possibility of building both a long-range and a medium-range supersonic passenger plane—an undertaking championed by Sir Arnold Hall, director of the Royal Aircraft Establishment. The Supersonic Transport Aircraft Committee (STAC), as the group came to be known, was chaired by the highly respected and politically savvy Sir Morien Morgan, a brilliant RAE aeronautical engineer who later would be called the Father of the Concorde for his commitment to producing a supersonic aircraft. With Morgan at its helm, the committee conducted its investigations over a three-year period. Its list of recommendations, which led to feasibility studies by the British Ministry of Defence, called for building a supersonic plane that could carry 150 passengers between London and New York and cruise at Mach 1.8, or 2,222 kilometers per hour (1,380 mph), more than doubling the speed of subsonic planes, which generally cruise at just above Mach 0.8, or 987 kilometers per hour (614 mph).

The committee's recommendations arrived at a time when the political climate favored an aggressive new project, since British aircraft manufacturers had failed to achieve the commercial success of their American counterparts. During this dynamic period, the Americans and Soviets also were investing huge amounts of resources in space exploration. The space race had begun, and both countries were sending satellites and rocket-powered spacecraft into orbit, using some of the same technology developed to launch nuclear missiles. The proposal that the United Kingdom should be the first nation to design and build supersonic transport for the civilian market was an idea whose time had come. "By 1962," Christopher Orlebar writes, "Britain had convinced herself that it was both highly desirable and technically possible to build an Atlantic range (3,700 statute miles) SST."

Such an ambitious project, if successful, could generate both prestige and profits. But it would come with a price tag that would require the full financial backing of at least one government, and preferably two. Britain needed a suitable, well-funded, and technologically sophisticated European partner. At that time, France was zooming into the future with a series of supersonic test aircraft. The star of the Paris Air Show of 1961 was the Super Caravelle, a model of a supersonic passenger plane and a product of Sud Aviation, the French state-owned aircraft manufacturer.

The Super Caravelle was similar—and a potential rival—to a British supersonic passenger plane, the BAC 223, or Bristol Type 223, that then existed only on paper in concepts developed by designers for the Bristol Aeroplane Company, which in 1959 merged with other major aircraft companies to form the British Aircraft Corporation (BAC). Perhaps the most advanced feature of the BAC 223 was its thin delta wing, the product of work being done at the RAE by German scientists and mathematicians, notably Johanna Weber, whose groundbreaking research into extended delta wings had enabled British aerodynamic experts to understand the phenomenon of vortex lift, a complex and previously mysterious source of aerodynamic lift arising from swirling eddies of airflow over aircraft wings.

Rather than compete to be the first to build a supersonic plane for the commercial market, Britain and France began engaging in talks. In November 1962, representatives from both countries signed an agreement, registered at the Hague, that bound them together in a shared endeavor to fund and develop a supersonic passenger plane; expenses and design responsibilities were to be shared equally. Sud Aviation and its successor, Aérospatiale, would take the lead in developing the airframe, with support from the British. Engineers working for BAC and British Airways would oversee the engine design, with help from the Société Nationale d'Étude et de Construction de Moteurs d'Aviation, or SNECMA, an aerospace engine manufacturing concern headquartered in the Paris suburb of Courcouronnes. (Design of the automatic flight control system would later be subcontracted by Aérospatiale to a British defense contractor.) The two countries would split the profits resulting from sales. Plans called for two prototypes to be completed and flown in four years, with production to follow by the end of 1968. Development costs were projected to add up to £170 million. The final cost would soar past £150 billion.

Once the British-French consortium had been formed to proceed with the Concorde project, the challenge that lay ahead was clear: to be the first to design and build a commercially viable aircraft capable of carrying 100 or more passengers at supersonic speeds. The terms of the agreement were straightforward. But the engineers who took on the task of designing the revolutionary aircraft had the additional burden of ensuring that their creation was not simply incredibly fast but also guaranteed to be thoroughly safe, comfortable, and passenger-friendly. It was one thing for test pilots equipped with oxygen masks and pressure suits to fly their supersonic aircraft at 16,764 meters (55,000 ft) in the thin air and freezing temperatures that prevail at such altitudes. It would be quite another matter to build a plane in which it was possible for passengers to lean back and listen to soothing music, unaware that the exterior of the fuselage in which they traveled was being heated to the boiling point of water.

The plane that would bear the name Concorde would require special heat-resistant aluminum alloys and high-reflectivity white paint for its skin and surfaces, and it would need unique engineering strategies for its wings, airframe, engines, and operating systems, as well as for carrying huge amounts of fuel. Lurking in the minds of its designers was the very real possibility that even the best-built plane, once in the air, would demonstrate unexpected mechanical and operational problems, or what Orlebar describes as "totally unforeseen side effects that could take years to iron out." Finally, civil aviation authorities who would be asked to certify Concorde as airworthy would have to write new rules to satisfy performance and safety requirements. "Every detail of the new aircraft," writes Orlebar, "would suffer the expert scrutiny of the civil aviation authorities of Britain, France and, as a prospective purchaser, the United States."

"It is not unreasonable to look upon the Concorde as a miracle," wrote the late test pilot Brian Trubshaw, who died in 2001. Because he was involved in every phase of the development and testing of

Concorde, Trubshaw was more than qualified to comment in hindsight on the scope of the undertaking and the myriad difficulties encountered and overcome along the way. "Who would have predicted that the combination of two governments, two airframe companies, two engine companies—each with different cultures, languages, and units of measurements—would have produced a technical achievement the size of the Concorde?" A fundamental disagreement arose at the planning stage about whether to design and build one long-range plane—an approach favored by the British—or both a long-range *and* a medium-range plane, which is what the French wanted and fought for but failed to achieve.

Avoiding misunderstandings due to language and cultural differences proved daunting. Over time, the French project leaders showed themselves to be more adept at learning English than the other way around. Rivalry and distrust threatened to undermine the underlying spirit of cooperation. "Nationalism was present on both sides and each party harbored suspicions that the other was trying to do them down," observed Trubshaw. Endless practical difficulties lay in wait as well, not least of them the clash between the imperial and metric systems. In England, measurements still were made in inches and feet, while the French did their work in centimeters and meters, presenting ongoing opportunities for errors whenever conversions were required so that parts would match and fit together. There were arguments between French and English pilots about matters large and small, including visual displays on the instrument panel. French pilot consultant André Türcat won his battle to include a "moving map display" in place of the conventional gauge layout on the instrument panel in the cockpit, a change Trubshaw strongly opposed, writing, "It was a strange philosophy for an aircraft that was going to spend most of its time over the sea." Türcat prevailed, but in the end, "the airlines chucked it out," said Trubshaw, "replacing it with a conventional instrument layout using gauges identical to the rest of their fleets."

Other aircraft makers producing high-speed bombers after the war already had solved some of the technical difficulties that confronted the

Concorde teams in both England and France. Chief among those were turbojet engines with afterburners, such as those powering the Convair B-58 Hustler, an American supersonic bomber that took its first flight in November 1956. Its four General Electric J79-GE-1 turbojet engines with afterburners gave it tremendous speed and earned it the nickname "Greased Lightning."

Some 250 engineers for British Airways helped to design the airframe and its engines. The British Concorde team quickly settled on engines manufactured for one of their own high-altitude bombers, the Avro Vulcan, flown by the Royal Air Force since 1956. The Avro Vulcan was not as fast as the Hustler, but its Rolls-Royce Olympus engines had proven effective on long-range missions and later evolved to include the necessary afterburners Concorde would need to get off the ground. Just as important, Rolls-Royce was a venerable, highly respected firm that had been designing and building engines for British planes since the start of World War I.

The aerodynamic challenges of supersonic flight facing the Concorde team were more problematic. There were no ready solutions at hand to adopt from those devised by makers of military aircraft because the requirements of military planes were different, starting with the amount of time that Concorde would be required to spend flying at supersonic speeds. Surprisingly, military planes do not spend much time flying at speeds above Mach 1. They accelerate to supersonic speeds only in special circumstances, such as when they must evade enemy missiles. The vast majority of the time, military jets cruise along at subsonic speeds. By contrast, Concorde would travel at subsonic speeds only during takeoff and landing, spending almost all its flying time at supersonic speeds. This difference would later be illustrated when the tiny Concorde fleet would rack up more hours of supersonic flight during the life of the Concorde project than the sum total of all the military planes in the world combined.

For designers, the time spent at subsonic versus supersonic flight was fundamental, because the aerodynamic requirements for subsonic flight are significantly different from those for supersonic flight.

Designs that work efficiently at subsonic speeds do not perform well at supersonic speeds. So compromises had to be made. Concorde had to be designed primarily for supersonic flight, with adjustments added for its short periods of subsonic flight during takeoff and landing. Nothing like this had ever been attempted before.

Foremost among the difficulties confronting the Concorde engineers was wing design. Shock waves encountered at transonic and supersonic speeds create too much drag and destroy the lift forces of ordinary wings, which allow planes to take off and land at subsonic speeds. In the early 1960s, there were two known solutions to the problem: swept-back wings and thin delta wings. The most efficient theoretical design for Concorde—the thin delta—had not been widely adopted by the military because of the design's inability to provide sufficient lift for takeoff and landing. Instead, the military opted for swept-back designs, even though they were not as efficient during the brief periods its jets spent at supersonic speeds.

This answer to the dilemma—the tradeoff between supersonic efficiency and subsonic lift—came from German mathematician Johanna Weber and her colleague Dietrich Küchemann. They had met and begun working together at the Aerodynamische Versuchsanstalt, or Experimental Aerodynamics Institute, in Göttingen, Germany. During World War II, both scientists were dissidents politically opposed to the Nazi Party. After the war, first Küchemann and later Weber emigrated to Britain and began to conduct research for the RAE in Farnborough, Hampshire. Weber, swept up by Operation Surgeon—a secret intelligence-gathering program responsible for bringing German scientists to Britain to prevent the Soviet Union from exploiting German expertise that could be used to build long-range bombers—became an RAE aeronautical expert.

While British forces were putting Weber's mathematical mind to work for the RAE, the Americans were rounding up German V-2 rocket scientists conducting research under the guidance of physicist Wernher von Braun, who later would put the first American satellite into orbit. However, it was Weber's mathematical theories that would put the first passengers on a British-French SST while American astronauts were landing on the moon.

Upon her arrival at the RAE, Weber was assigned to work with British researcher John Seddon, who shared her in-depth knowledge of propulsion dynamics. "Küchemann had evidently sung her praises before her arrival," observes John Green, past president of the Royal Aeronautical Society, "with the result that Seddon greeted her with 'Oh, the Myth has become a Miss.'" Weber was mentored by the highly respected Frances Bradfield—known to her colleagues as Miss B.—who was then a supervisor in the Aerodynamics Department of the RAE. Bradfield was a mathematician, a graduate of Newnham College, Cambridge, and "an exacting but kindly boss," according to Green. Bradfield took it upon herself to look after the famously shy Weber, who lived alone in a small one-bedroom apartment and sent most of her earnings back to Germany to support her elderly mother and younger sister.

Weber's work was aided by a team composed of women technicians, known as "computers," who performed calculations by hand, freeing Weber to pursue theoretical investigations. Weber's work was responsible for the successful wing design of the Handley Page Victor, a high-speed subsonic strategic bomber introduced in 1958. After that success, Küchemann and Weber worked together to improve the theoretical methods then being applied to subsonic aerodynamics. Determining the amount of lift that a wing will create is complicated and involves many factors, including the key features of wing length, thickness, camber, twist, and sweepback. Aerodynamicists, including Weber, had modeled each factor separately. However, it was Weber who first managed to combine all the factors into a single mathematical model that treated them simultaneously. This simultaneous treatment was critical because in actual practice, the various factors interacted with one another and dramatically affected the wing's real-world performance. Prior to Weber's mathematical models, the only way to simultaneously determine wing lift was by constructing physical models and undertaking time-consuming testing in wind tunnels.

Her calculations showed that a radical thin delta wing, extending far down the length of a plane, could provide the lift needed to allow it to take off and land at subsonic speeds yet also maintain efficiency at supersonic speeds. The equations she formulated predicted that a significant increase in vortex lift over the extended delta wing at subsonic speeds would get the plane off the ground. She was right. Weber's published papers on the vortex lift of extended delta wings radically altered the entire understanding of supersonic flight, and many of those involved in the early stages of supersonic research considered her ideas to be both foundational and pivotal.

Despite her achievements, Weber relied on Küchemann to present her findings at meetings and public forums, which kept her in the background. "There is no doubt in my mind that Küchemann's presentation in favor of the slender wing concept was the inspiration for the wing configuration of Concorde, and Morien Morgan [then chairman of STAC] locked on to this immediately," observes RAE test pilot and author Eric Brown. The resulting wing designed by the Concorde project team—known as an ogival thin wing because of its similarity to a pointed Gothic, or ogive, arch—was not only a technological breakthrough but also an aesthetic triumph that produced one of the most beautiful aircraft ever to fly.

Weber's work also led to the second most visually obvious feature of Concorde: its hinged needle nose, or "droop snoot." The thin delta wing design required an extreme angle of attack to develop Weber's predicted vortex lift at subsonic speeds. The angle was so steep that the nose of Concorde would obscure the sightlines of the pilots during takeoff and landing. The engineering solution was to hinge Concorde's nose so that it could be dropped out of the way during subsonic flight and then rotated back up when the plane reached supersonic speeds, where vortex lift was no longer required.

While the thin delta-wing design solved the subsonic and supersonic lift dilemma, there remained a second serious design problem: devising a control system for Concorde that could function at both supersonic and subsonic speeds. Pilots of subsonic aircraft use their

Concorde was characterized not only by its delta wing but also by its hinged needle nose, or "droop snoot." Concorde took off at such a steep angle that it required a nose that could be moved downward, out of the pilots' line of vision, and then lifted into place again once the plane reached supersonic speeds. (© STEVE FITZGERALD)

planes' control surfaces, such as ailerons and flaps, to create aerodynamic forces that change the planes' direction. To change the pitch of an aircraft in order to climb, a pilot elevates the horizontal stabilizers at the tail of the plane. When the oncoming airstream impacts the raised stabilizers, a downward force is exerted on the tail; the front of the plane then pivots up around the center of gravity, and the plane angles upward and climbs.

While this system of pitch control works well at subsonic speeds, it is impractical at supersonic speeds. The resulting drag forces on the stabilizers from the blast of the supersonic airstream would simply be too great for the aircraft to overcome. So Concorde's engineers devised a different system to control pitch, one that would work at both subsonic and supersonic speeds. It involved using fuel to shift the center of gravity of the aircraft. Fuel was pumped back and forth between pairs of what are called trim tanks—tanks located at the plane's nose and

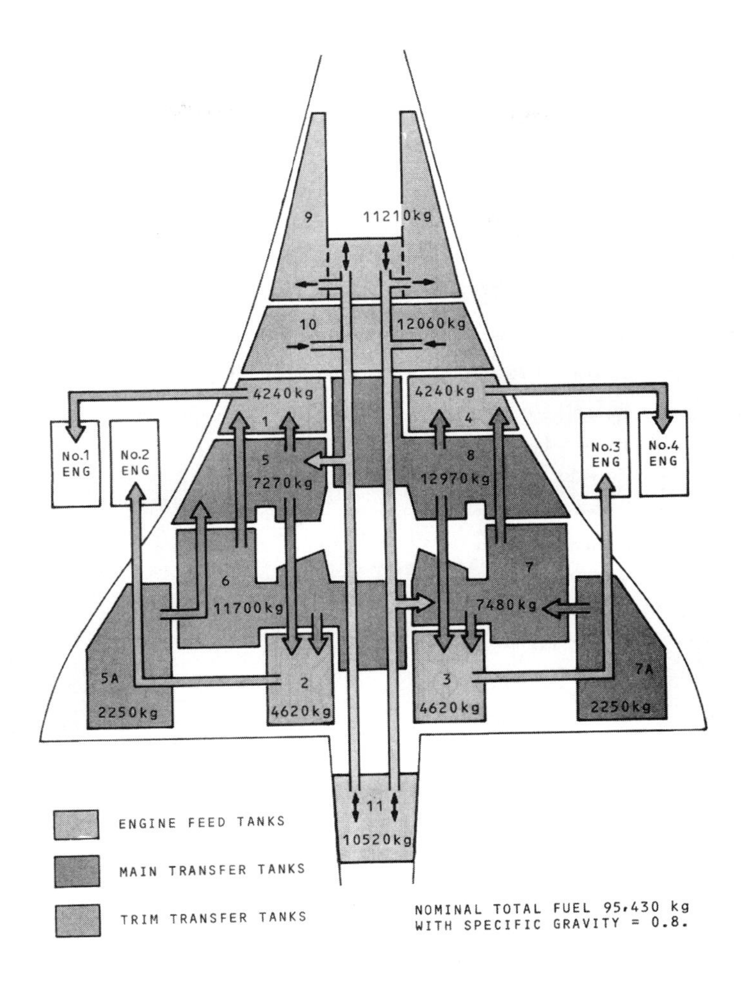

AF BA

Concorde's complex system of 13 fuel tanks, set in the fuselage, wings, and tail. To maintain the plane's center of gravity and keep the jet level in flight, fuel was pumped continuously among the tanks, with the flight engineer monitoring its movements. Concorde could consume as much as 100 tons of fuel on a transatlantic crossing, a massive expense that factored into the jet's eventual retirement. (DAVID LENEY AND DAVID MACDONALD, *AÉROSPATIALE / BAC CONCORDE, 1969 ONWARDS [ALL MODELS]: OWNERS' WORKSHOP MANUAL* [SOMERSET: HAYNES PUBLISHING, 2010]) tail and used to control the aircraft's trim or pitch. This solution made ingenious use of the staggering amount of fuel Concorde would need to fly across the Atlantic Ocean.

Pilots could also use the weight of the fuel to help the plane get off the ground. Taking off at subsonic speeds required that the nose of the aircraft be positioned at a higher angle of attack. Prior to takeoff, pilots could pump fuel into a trim tank near the tail of the plane. After reaching cruising altitude, the fuel would be pumped out of the aft trim tank ahead to the forward trim tank. This would shift weight forward and pull the nose lower to reduce the pitch of the plane to the flatter attitude required for efficient flight at supersonic speed.

Concorde's supersonic design necessitated two additional design compromises at subsonic speeds: extendable landing gear and special tires. The extendable landing-gear system on Concorde was more complicated than the gear on conventional aircraft. At rest at the gate, when Concorde's fuselage was level to the ground, the shortened landing gear looked similar to the gear on other aircraft. However, prior to takeoff, the landing gear had to be extended significantly to raise the nose to the appropriate attack angle required for takeoff. Once airborne, the long, extended landing gear hanging down into the airstream created significant drag, so it was vital that the landing gear be retracted quickly in order to gain altitude.

When an aircraft is at rest on the runway prior to takeoff, the only force on the tires is the gravitational force generated by the mass of the plane: its weight. However, as the plane begins to accelerate down the runway, two new forces come into play, centrifugal force and aerodynamic force. On a subsonic plane, air flowing across the wings generates upward aerodynamic lift forces that oppose the downward force of gravity on the plane as it accelerates, resulting in steadily reduced downward pressure on the tires. But this was not the case on Concorde. The thin delta wing did not provide equivalent aerodynamic lift forces at subsonic takeoff speeds, and thus the tires had to continue to bear the entire weight of the plane on takeoff. Concorde's higher takeoff speeds prior to rotation also meant that its tires would be subjected to higher centrifugal forces that could cause them to fly apart in the case of a blowout.

Taking this into account, Concorde engineers required that their tire suppliers meet more stringent specifications for customized, durable tires that could withstand these more intense forces. However, over time, even these specially designed tires would experience higher rates of failure than would tires on similarly sized subsonic planes. Ultimately, the vulnerability of Concorde's tires to failure and the plane's difficulty in gaining altitude with extended landing gear both played a significant role in the crash of Concorde at Gonesse.

•

After a range of design strategies had been agreed upon by the English and French engineers, hundreds of parts and working components were manufactured for assembly in Filton, in southwest England, and Toulouse, in the south of France, which was the hub of France's vibrant aeronautics industry. Some of the first Concordes that came off the assembly lines would be used for testing. A special facility was constructed at the RAE complex in Farnborough, where an entire Concorde fuselage, with its single wing, was housed and used as a "fatigue specimen," heated and cooled to simulate the wear and tear of flight cycles from takeoff to landing. An entire Concorde was submitted to strength tests at a government research center in Toulouse.

In all, 10 Concordes would be assembled at Filton and 10 more in the hangars at Toulouse. Of the 20 Concordes built, six would never be sold. Four of those six were preproduction models used in a variety of tests. The other two, numbered 001 and 002, would be the all-important prototypes used for inaugural test flights. One of the six preproduction planes would eventually be scrapped, according to Trubshaw, while the rest would be retired to museums, including prototype 001, which would be donated to the Musée de l'Air et de l'Espace, or Museum of Air and Space, at Le Bourget Airport. "It always seemed a great pity," wrote Trubshaw, "that there was no further use for these

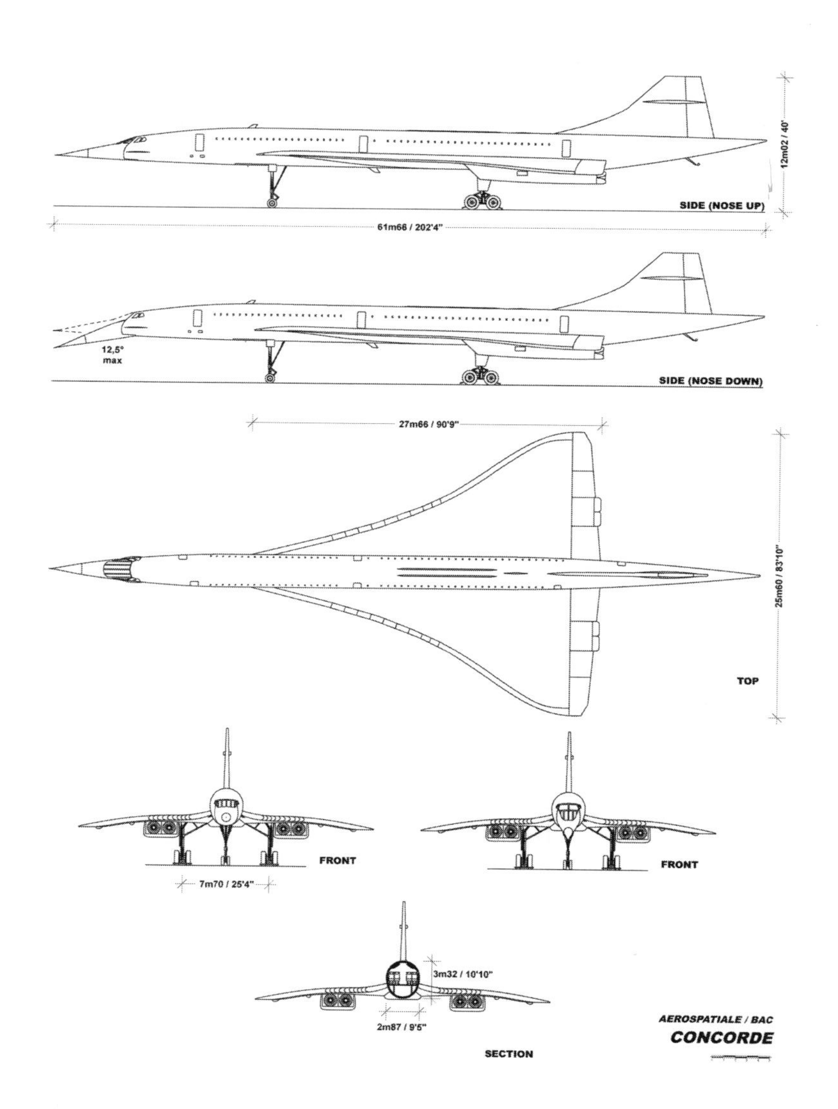

General plan of Concorde as it was manufactured by Aérospatiale/BAC: views from the side, the top, the front, and in section. (*BEA REPORT*, APPENDIX 1)

aircraft, which had been the backbone of the development program and which had flown few hours."

Before 001 and 002 made their debut, researchers and pilots spent hundreds of hours in specially constructed, fully enclosed flight simulators mounted atop large mechanized platforms that replicated the conditions of actual flight. "The simulator was not entirely reliable and was known to run away and jam against the wall of the building where it was situated," according to Trubshaw. "A long wait followed, often in somewhat hot and sticky conditions, while a ladder was produced to extract the crew."

On March 2, 1969, the French were the first to fly their prototype, number 001, from a rural airfield in Toulouse. The much-awaited event took place after the engineers had spent two years test-flying the Anglo-French Concorde on a simulator, plus two days spent waiting for suitable weather conditions; 48 hours of strong winds and the threat of rain delayed the maiden flight. At the controls during the 28-minute subsonic roundtrip to and from the airfield would be Sud Aviation test pilot and engineer André Türcat, described as tall, jocular, and meticulous by the British press. "The engineer's aim is to remove all but the faintest traces of novelty and adventure from the maiden flight," wrote David Fairhall of the *Guardian*. "In the prosaic language of the test pilot, the object of the exercise is simply to check the accuracy of the simulator in which he has been flying the aircraft on the ground for the past two years." Türcat was "fastidiously calm" despite the clouds and gusting winds; while he waited for the skies to clear, he lay down for an after-lunch nap.

When Concorde finally did take off, there was enough drama to satisfy those in attendance. The plane reared up for takeoff, in Fairhall's words, "like some monstrous swan," as the roar of its afterburners "silenced the airfield larks and scattered the starlings." After a few modest maneuvers, the plane successfully touched down on its first approach. When he emerged from the cockpit to the sound of applause, Türcat announced in both French and English: "The big bird flies."

Concorde's maiden flight: prototype number 001, piloted by Sud Aviation test pilot and engineer André Türcat, takes off from a rural airport in Toulouse, France, on March 2, 1969. The British would follow with the first flight of their own prototype, 002, just over a month later. (FONDS ANDRÉ CROS, CITY ARCHIVES, TOULOUSE, FRANCE)

Just over a month later, on April 9, another successful test flight was completed under clear skies by Concorde prototype 002 over the Severn Valley near Bristol. Despite two malfunctioning radio altimeters and a balky afterburner, the flight went smoothly. Trubshaw was its pilot, and he flew 56 uneventful kilometers (35 mi) from Filton Airfield to a Royal Air Force Air Support Command Station near Fairford, where he had to circle the airfield to avoid a small private plane that made an unauthorized appearance just as he was preparing to land. He emerged from the cockpit in high spirits, however, declaring of the flight: "It was wizard—a cool, calm, and collected operation." Yet another Concorde would pass its first supersonic test flight on October 1, 1969.

•

As the new planes, intended to be sold to major airlines, were taking shape in hangars in Filton and Toulouse during the 1970s, hopes were high for sales of more than 200 Concordes, despite the threat posed by emergent competitive supersonic programs in both the United States and the Soviet Union. But the political winds did not favor an American government-funded SST, which was abandoned in 1971 after being assailed by its critics, including then–Secretary of Defense Robert McNamara, as a costly project that would benefit only wealthy jet-setters and therefore would be a poor use of taxpayer money.

The Soviet Union pressed ahead with its own version of Concorde, however: the Tupolev Tu-144. After making its maiden flight in December 1968, it exceeded Mach 1 for the first time in June 1969. It was a fine launch for a plane that, sadly, never lived up to its promise. The Tupolev Tu-144 instead became notorious for a spectacular crash at the Paris Air Show in 1973; what had begun as a dazzling display of the plane's ability to perform aerobatic twists and turns ended tragically when it broke up and crashed into the village of Goussainville. The six people onboard the plane were killed, as were eight on the ground. (A heated but never-resolved controversy arose when it was learned that the Soviet pilots had never been told that they would be sharing the airspace with a French airplane sent aloft to photograph the Tu-144, which reportedly was forced to take evasive maneuvers to avoid a midair collision.) After another crash in 1978 and a brief run carrying passengers inside the Soviet Union, the Tu-144 was withdrawn from service and ignominiously converted to a cargo plane; it was never operated commercially outside the country.

The achievements of the Concorde design teams in Britain and France would be followed by dynamic production and assembly work by BAC in Filton and Sud Aviation in Toulouse. Yet finding customers would prove problematic, since a supersonic plane presented significant problems in terms of public acceptance. Concerns were raised about whether passengers would be exposed to unhealthy levels of cosmic radiation at high altitudes. Environmentalists argued that nitrous oxides in Concorde's exhaust could rise upward and deplete the protective ozone layer above the earth. And sonic booms—those thunderous claps that are unavoidable by-products of supersonic flight—were chief

The Soviet Union also built a supersonic passenger jet, the Tupolev Tu-144, but it was soon relegated to cargo transport. Here a later experimental version, the Tupolev Tu-144LL supersonic flying laboratory (built in a collaboration between NASA and the Russian aerospace company Tupolev ANTK), takes off from Zhukovsky Air Development Center, near Moscow, on a 1997 test flight. (NASA / IBP)

among the drawbacks of introducing a supersonic jetliner that would fly over populated regions and land at airports located in or near large metropolitan areas. Some strategies would be devised to lessen the impact of the roar generated by the engines and afterburners during takeoff, but nothing could be done about the booms. A sonic boom results when an aircraft exceeds the speed of sound and generates powerful shock waves that are both heard and felt on the ground. A common misconception is that the boom is produced only at the moment when an aircraft surpasses Mach 1. In reality, the noise is continuous once a plane surpasses the speed of sound. As Trubshaw explained. "If you could run as fast as the aircraft, you would hear a sonic boom all the time, creating a sort of carpet."

The 1970s would prove disappointing in terms of the plane's prospects, which were negatively impacted by changing market forces and Concorde's intrinsic twin liabilities: high fuel consumption and noise pollution. The oil crisis of 1973 prompted major carriers from Asia to the United States to shun Concorde in favor of slower but more fuel-efficient competitors, and the daunting problem of the plane's noise, both its loud takeoff roar and its sonic boom, could not be solved. The fact that Concorde was not only incredibly fast but also incredibly noisy meant that the push to establish routes in the United States met with widespread and intense coast-to-coast protests by people living near airports. Grassroots organizations sprang up, and members of Congress applied political pressure. As a result, no Concorde would ever fly overland in the United States, from New York to Los Angeles or between any other hubs; these lucrative transcontinental routes were denied to it, with predictable results. The hoped-for sales of more than 200 Concordes to major airlines never materialized. The problem of the sonic boom "figured highly in sales deliberations," says Trubshaw, "because of its potential to limit routes, although there were a number of areas [outside the United States] where supersonic flight overland was seen to be possible, such as Iran, Saudi Arabia, and Australia."

Although Concorde was an engineering triumph, it was proving to be a commercial flop. TWA and Pan American Airlines cancelled their options to buy Concorde, which also failed to win converts among major airlines in Europe and Asia. Marketing efforts were made to sell Concorde to Philippine Airlines, including flying Madame Imelda Marcos in a Concorde to Hong Kong for a shopping trip. "The trip only took one hour, as there were no supersonic restrictions," wrote Trubshaw. "The only problem was finding room on the aircraft for Madame Marcos's purchases." When one member of the French sales team asked Madame Marcos what she would like to do the next day, she enthusiastically answered that she wanted to return to Hong Kong for more shopping. Despite the red carpet treatment, however, "nothing materialized" in terms of sales of Concordes to Philippine Airlines.

The only remaining customers were British Airways and Air France. In July 1972, British Airways bought five Concordes, valued at £164 million, for the bargain price of £13.2 million per aircraft (for a total of $66.3 million); another two were essentially given away to British Airways for the token sum of £1 apiece. Air France signed a contract at the same time to buy four Concordes at the same price, followed by individual sales of three more planes over the next eight years. Concorde F-BTSC, for instance, was leased to Air France in 1976 for daily flights until other aircraft were delivered. It was returned to Aérospatiale later that same year and placed in storage. Air France leased it once more in 1979 before it at last purchased the plane for one franc in October 1980.

Air France proudly put its livery on its small fleet and prepared to sell tickets. One French newspaper ran a cartoon of a champagne glass held up for a toast, with the caption "We have it all to ourselves." The British press, however, was indignant at the general rejection of the great plane and blamed American protectionism and "jealousy" for Concorde's failure to win widespread acceptance in the United States.

•

Commercial service began with considerable fanfare on the morning of January 21, 1976, when two Concordes took off in synchronized departures: an Air France Concorde flew out of Paris to Bahrain in the Persian Gulf, and a British Airways Concorde left London for Rio de Janeiro, Brazil. Most of the seats on both jets were filled with dignitaries from France and England, of course, as well as notables from Germany, Sweden, Brazil, and Spain. Queen Elizabeth II sent her "warmest congratulations" to the people of France for the "successful outcome of 14 years of close collaboration between our two nations."

Over ensuing years, many well-heeled passengers came forward to purchase tickets to fly from London or Paris to New York and back in one day. Although some flights took off only half full, Concorde was to remain a popular choice for dignitaries and celebrities. Pope John Paul II flew on Concorde, as did Queen Elizabeth, Prince Philip,

Queen Elizabeth II and Prince Philip disembark from a British Airways Concorde on May 20, 1991. Concorde's reputation for glamour was built in part on its passenger lists of royalty, business titans, and celebrities, among the few who could afford its expensive fares. (SRA JERRY WILSON)

and French President Jacques Chirac. Celebrity passengers included Elizabeth Taylor, Sean Connery, Robert Redford, Madonna, Diana Ross, Luciano Pavarotti, the Beatles, and the Rolling Stones. Sir David Frost, the famed British television host, was a fan of Concorde and claimed to have flown it between London and New York more than 300 times. Charter Concorde business proved to be brisk, even if general commercial service was not: Goodwood Travel and Superlative Travel in the United Kingdom arranged specialty flights on Concorde for travel experiences that included spending Christmas above the Arctic Circle in Scandinavia and supersonic jaunts to Moscow to attend performances of the Bolshoi Ballet. In August 1995, Concorde 210, owned by British Airways, flew around the world on a 23-day air cruise, making stops in the Pacific islands, Beijing, India, and Africa.

The 13 Concordes in service continued to perform like powerful, pampered thoroughbreds. The British Airways–owned Concorde 210 set a transatlantic record on February 7, 1996, during a routine flight from New York to London that took just two hours, 52 minutes, and 59 seconds. During the crossing, it averaged 35 kilometers (22 mi) per minute, prompting British Airways flight manager Mike Bannister to point out that the seven planes in BA's fleet of Concordes were in "great shape" and remained "sprightly and youthful."

One of the most interesting features of all six Air France Concordes was the electronic indicators in the cabin that displayed the changing Mach number, so that everyone on board could see the exact moment the plane surpassed the speed of sound, an event that was simultaneously announced by one of the pilots. Otherwise passengers typically reported feeling only a slight surge in acceleration as Concorde achieved its previously unimaginable speeds.

On July 25, 2000, the German tourists who settled into their leather seats on the chartered Concorde 203 had every reason to be excited about their upcoming adventure and to believe that flying on an Air France Concorde was even more safe than flying on a subsonic plane operated by a less prestigious airline. Concorde had an unblemished record in this regard, having never had a single accident resulting in a fatality. After all, every effort had been made to keep both the French and the British fleets well maintained and their flight crews trained to the highest standards.

The world's first supersonic passenger planes had exceeded the expectations of the Anglo-French consortium that conceived and built them over a 14-year period. Concorde was a dream made manifest, a plane whose makers had conquered the daunting design challenges of sustained supersonic flight. In the year 2000, its saga was still ongoing, however, and elements of both irony and tragedy would come into play. Concorde F-BTSC, chartered to fly from Paris to New York, would not fail as it was screaming across the ocean at 18,288 meters (60,000 ft) at Mach 2. Instead, it would meet a seemingly more prosaic fate, catching fire while it was still on the runway and traveling at less than 200 knots (370 kmh or 230 mph). It would be up to an international team of investigators to determine how and why that had happened.

5

A Trail of Evidence

The first light of morning on Wednesday, July 26, 2000, did little to dispel the gloom over Paris or lift the spirits of investigators assigned to unravel the mystery of Flight 4590. The battered remnants of the Air France Concorde lay scattered on sodden ground blackened by oily soot. The entire zone remained cordoned off with police tape, creating the impression of a large crime scene as more bodies were located and removed on stretchers carried by weary firefighters.

German and French relatives of the 100 passengers and nine crew members onboard were reeling from their loss and trying to understand how a routine charter flight could have gone so terribly wrong. Among them was the family of the lone American on the plane: Christopher Behrens, 65, a native of New Jersey who lived in Frankfurt and had worked for Air France. Behrens had "loved flying the Concorde," his older sister, Rachel Coates, says. "His job required him to fly it often. He flew the Concorde around the world four times. He loved how fast it traveled." The unmarried Behrens, who had recently retired, "loved life" and travel, just as he had enjoyed his Air France career, Coates says.

On day two of the investigation into the crash of Flight 4590, the international team of air safety experts knew that arriving at a conclusion that would establish the cause of the crash with certainty would be difficult. "The philosophy of air safety investigation is probable cause," says Bob MacIntosh, an investigator for the National Transportation

Safety Board who took part in the French-led investigative effort. "We're not in it to do the absolute proving, as you might be required to do in a criminal case. Our job is to make recommendations for improving aircraft safety and to prevent a recurrence of events that led to the crash. [We] go with probability rather than with absolute proof. That's pretty much where the BEA [Bureau d'Enquêtes et d'Analyses, the French aviation agency overseeing the investigation] had to go with the Concorde investigation."

Every crash of a passenger plane involves the abrupt and alarming intrusion of the unexpected into the risk-averse world of daily air travel. In most investigations, not one but many causal factors are found to have contributed to a major accident that resulted in loss of life and the destruction of an aircraft. It is not unusual for the most critical of those factors to be initially obscured by minor malfunctions and misleading clues. Still, MacIntosh and the other investigators were keen to find the definitive clue, or set of clues, that would solve the high-profile case—the first-ever crash of a supersonic Concorde.

The investigators working under the purview of the BEA would need all their combined years of experience to improve their chances of solving the puzzle presented in Gonesse. First they would painstakingly gather all the evidence, analyze it, and conduct tests, and then they would arrive at a forensic reconstruction of events that they hoped would credibly explain what had gone wrong and why. One early and fortunate discovery improved their odds of success: both of the plane's so-called black boxes—the cockpit voice recorder and the flight data recorder—had been salvaged from the smoking wreckage before the end of the first day, and they were found to be largely undamaged inside their two layers of protective insulation and steel.

Investigators on the scene first walked among the charred and mangled remains of the Concorde. Witnesses had seen the plane bank and roll over, but by the time it struck the Hotelissimo in Gonesse, it had leveled out and effectively pancaked. Its 13 fuel tanks, including the large main wing tanks, had been filled to the brim in preparation for the long transatlantic flight to New York, which contributed to the intensity of the initial explosion and ensuing fire. It had taken three hours to extinguish the blaze fed by tons of jet fuel and the remains of the two-story wooden hotel, which had been reduced to a ruin. The fire had been so intense, in fact, that plastic fixtures on the exterior of a neighboring hotel had melted. More than 180,000 liters (47,550 gallons) of water had been used to tame the inferno, along with 3,800 liters (1,004 gallons) of emulsifier applied by six vehicles equipped with foam firefighting systems.

The bodies of the three members of the cockpit crew, Captain Christian Marty, First Officer Jean Marcot, and Flight Engineer Gilles Jardinaud, were found still strapped into their seats inside the battered hull of the cockpit, which sat on its side on the southern edge of the large debris field. The biggest pieces of the wreckage, including segments of the delta wing and the four turbojet engines, which were 4 meters (13 ft) long and weighed 3,175 kilograms (7,000 lb) each, dwarfed the BEA technicians standing nearby.

Much smaller but more telling were the soot-covered pieces of debris that lay strewn along runway 26R, in the southeastern sector of Charles de Gaulle Airport, and on the ground along the plane's flight path. Neither would the investigators want for detailed accounts from witnesses, who had watched in disbelief as Concorde plummeted and burst into a fireball visible for miles. Instead, their task would be to consider the plentitude of evidence as a whole and puzzle out how the myriad pieces fit together and related to one another.

Before they began formulating hypotheses as to what occurred that Monday afternoon, Alain Bouillard and his crew of air safety detectives had to do their homework. That would involve using one of the most fundamental tools at their disposal: onsite observation. Bouillard and his investigators immediately organized airport workers to help them conduct an exhaustive inspection of the place where the terrible series of events had begun: runway 26R. "We became interested in the runway because the event happened during acceleration for takeoff," says Bouillard. "So one team from the BEA left really quickly to inspect runway 26R."

There they would survey every meter of the concrete surface, looking for debris scattered along the runway and beyond, since more fragments of the plane were found past the airport to the west, along a swath of ground beneath the flight path of the burning aircraft. But the main focus would be on the runway itself: "From witness interviews and so on, we knew that things started to go bad about 1,700 meters [5,577 ft] down the runway," says MacIntosh. "Consequently, the folks at Aéroport de Paris [Charles de Gaulle] were out on the runway the very first day, looking for debris, sweeping it up, and getting it mapped properly so that it could be forensically understood as to what dropped first and what interactions took place on the runway."

When every stray fragment found on the runway and its environs had been collected and identified, the BEA investigators would do their best to step back and consider as a whole the assortment of objects, as well as important trace evidence on the runway itself, including a large fuel stain, blotchy soot deposits, and intermittent tire tracks. Some of the investigators were assigned to walk the runway and try to interpret any skid marks that might have been left by Concorde's tires. Their goal was to construct an event-driven narrative of what had occurred during the failed takeoff and short flight.

Air traffic controller Gilles Logelin had been monitoring the plane's progress from the southern control tower very near runway 26R when he spotted flames streaming from the underside and rear of the plane and immediately informed the cockpit crew, who were unaware that their aircraft was on fire. Some witnesses interviewed by the BEA said they saw a luminous flash that preceded the appearance of flames. Somehow, the plane had begun releasing a large quantity of jet fuel, later estimated to be about 60 kilograms per second (3,600 kg/min or 7,936 lb/min), which had instantly ignited.

The inspection of the runway resembled a deadly serious version of a game of scavenger hunt as workers found and collected a strange assortment of objects. The east-to-west runway was 45 meters (148 ft) wide and 4,215 meters (13,828 ft) long, including 600 meters (1,968 ft) of asphalt tarmac joined to 3,615 meters (11,860 ft) of concrete. To aid in their analysis, the team divided the runway, 4.2 kilometers (2.6 mi) long, into 482 sections composed of the square concrete slabs of which most of the runway was constructed and which each measured 7.5 by 7.5 meters (24.6 by 24.6 ft).

Every object that was found and picked up was also identified by its location, since the runway represented a linear timeline; where things were found corresponded to when something happened in the lead-up to the crash. When the airport workers began collecting debris, they soon came across shattered pieces of Concorde's fiberglass water deflector plates, called cow catchers, that are mounted on the front of each of the two left- and right-main landing gear bogies to shield the plane's engine intakes from sprays of water shooting up from ground puddles or rain kicked up by the wheels during takeoffs and landings.

The fiberglass pieces of the water deflector were found scattered along the runway from slab 139 to slab 166—a distance of 202 meters (662 ft). In the same area, on slab 152, they discovered a large 4.5-kilogram (9.9-lb) piece of torn tread from one of Concorde's tires that was lying about 1,750 meters (5,741 ft) from the runway threshold. Farther along to the west were other, smaller pieces of tire tread, including one piece whose torn edge aligned perfectly with the damaged edge of the 4.5-kilogram piece, which bore evidence of a curved 32-centimeter (12.5-in) cut. "We found many elements of the plane, parts from the water deflector, of course, but we also found pieces of tire of different sizes," says Bouillard.

At the outset, one of the most pressing questions was how the plane had caught fire. To arrive at the answer, it would be necessary to first document the soot trail left by the flames streaming from the rear of Concorde as it traveled from east to west, picking up speed as it went. BEA investigators began studying other, highly visible trace evidence in the form of a large dark stain centered at 1,820 meters (5,971 ft)—the oily residue created by a splatter of unburned kerosene that appeared to have come from one of the tanks on the port side of Concorde's large delta wing. The stain appeared just 70 meters (229 ft) beyond the place where the 4.5-kilogram (9.9-lb) piece of tire debris was found. It

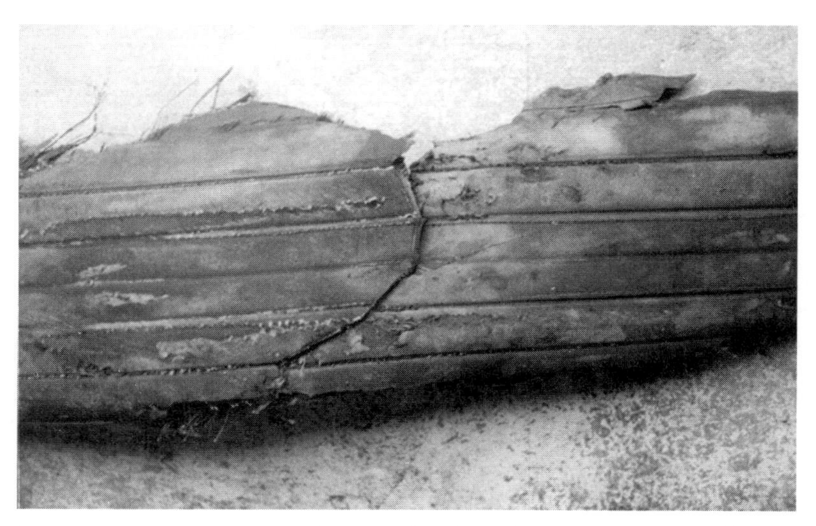

Found on runway 26R at Charles de Gaulle after the crash, this 4.5-kilogram (9.9-lb) fragment of Concorde's number 2 tire, punctured by debris left behind by another departing jet, had hurtled into the number 5 wing tank, causing an internal shock wave that ruptured the tank. (*BEA REPORT*, 60)

certainly looked as if whatever happened to the tire, which had been destroyed, had occurred before the fuel spill. The question was how the two were related.

Immediately after the fuel stain, investigators came across a trail of large, blotchy soot deposits on the left side of the runway's center stripe—the most visible evidence of the intensity of the flames that had come from the left underside of the aircraft. The heavy soot deposits were about 7 meters (23 ft) wide and amounted to a shadowlike imprint of the accident in progress. At first, the soot trail paralleled the center stripe of the runway, but then it shifted direction, drifting steadily to the left, mirroring the leftward yaw or tack that witnesses said the plane itself had taken. The black deposits, so like Rorschach patterns, made it very clear that the flames streaming off the plane had been fed by fuel from one or more of Concorde's tanks, most likely one of the large portside wing tanks, which included tank number 5.

The plane had also run over a safety light on the left edge of the runway on slab 293, 2,800 meters (9,186 ft) from the threshold of the runway—an event that took place as Concorde's left engines lost thrust while the right engines continued roaring along at full power, forcing the plane off the runway to the left. Small pieces of the broken light were discovered near deep gouge marks in the ground, indicating that the light was likely broken by an exposed metal wheel rim or parts of the left-main landing gear, which could not be retracted, shortly before Concorde lifted off.

Investigators were already assembling a very crude timeline of what had happened. Well before the plane was halfway down the runway, it appeared that one of Concorde's tires had burst. Some 70 meters (230 ft) farther on, a fuel tank had been holed or ruptured, spilling a large quantity of kerosene that had quickly ignited, generating flames on the underside of Concorde that grew in intensity, leaving behind a trail of soot that paralleled the path of the burning plane. The grass along the left side of the runway between 2,902 and 3,165 meters (9,521 and 10,384 ft) had been scorched, providing evidence of flames shooting out from the underside and rear of Concorde after it became airborne. Multiple burn marks on the runway and nearby ground identified the places where flaming chunks of debris were dropped—pieces of the aircraft itself. But what had caused the rupture in the tank?

Investigators immediately noticed evidence that something significant had happened farther on, about two-thirds of the way down the runway. The first dark track left by a single tire began at slab 161, about 1,800 meters or 1.8 kilometers (5,905 ft) from the threshold of the runway where Concorde had initiated its takeoff run. This was not the ordinary continuous track of a normal tire but the incomplete track of a badly damaged tire. The story written on the runway by this track and others was of a normal takeoff down the center of the runway until the aircraft reached a point 1,800 meters (5,905 ft or 1.1 mi) from the runway threshold.

At slab 161, the number 2 tire on the left landing gear had suffered a blowout. The first incomplete track from this tire indicated that the aircraft then continued on a straight trajectory down the center of the runway for another 400 meters (1,312 ft) to a point 2,200 meters (7,217 ft) from the runway threshold as the flailing tire tore itself apart. At that point, the aircraft then began its directional shift to the left for another 140 meters (459 ft), with the blown tire continuing to disintegrate until there was nothing left of it. Then, with the number 2 tire gone, its intermittent track disappeared at a point 2,340 meters or 2.3 kilometers (7,677 ft or 1.5 mi) from the runway threshold. The left-main landing gear was down to three remaining tires, which had to bear the weight that was supposed to be spread among four.

Another set of irregular tire tracks appeared at the point where the three remaining left-main landing gear tires showed signs of the stress of the increased weight, measured in tons, bearing down on them as the aircraft continued sliding to the left. These irregular tracks continued up to the broken landing light on the left side of the runway. At that point, these irregular tracks became intermittent, and they disappeared 30 meters (98 ft) farther on, near the left edge of the runway at a point about 2,830 meters (9,285 ft) from the runway threshold, indicating that the aircraft had rotated and gone airborne shortly after striking the light.

Later, investigators would learn that airports workers had discovered a square 32-by-32-centimeter (12.5-by-12.5-in) piece of aluminum debris at 1,800 meters (5,905 ft)—just 20 meters (66 ft) in front of where the fuel stain appeared. It was later found to be a structural piece from the aluminum hull of fuel tank number 5. Because of where the fragment was found, investigators regarded it as an important piece of evidence, possibly the piece of the tank that corresponded with the hole from which all the fuel had spilled.

When investigators got a good look at the large fragment, they were not surprised that it showed no signs of exposure to flames or high heat, since the puncturing of the tank would have preceded the fire. But they were puzzled as to why the fragment showed no signs of the sort of damage that would result from a high-velocity object, such as a piece of tire or water deflector, striking the tank with enough force to punch a hole in it. Both sides of the fragment, inside and out, were in remarkably good condition. It didn't make any sense. Solving the mystery of what had happened to the fuel tank would prove one of their greatest challenges over the course of the investigation.

The grim trail of evidence continued long past the end of the runway and all the way to the crash site in Gonesse. The fire-ravaged plane had continued to disintegrate and melt, shedding pieces of its wing and tail as it flew along at about 60 meters (197 ft) above the ground with its albatross-like landing gear still extended. Most of what was found between the airport and the crash site came from the left side of the plane and included an anticollision light on the tail cone, fire-damaged panels from the left-side delta wing's lower surface, eight panels from the upper surface of the left wing, and fire-damaged structural parts from the aircraft's distinctive pointed tail cone. "Leading up to the crash site, many small pieces of metal, honeycomb components, pieces of riveted structure and parts of the rear fuselage were found," says Bouillard.

The runway survey had yielded a plethora of trace evidence and fragmented objects—what Bouillard later termed "an ensemble of debris." Nonetheless, all the larger objects, regardless of condition, were readily identifiable, with one notable exception: a bent strip of metal 43 centimeters (17 in) long and punctured by a series of irregularly drilled holes that looked like holes for cherry rivets—that is, aircraft rivets, or fasteners, named for Carl Cherry, whose invention revolutionized aircraft manufacturing. It was coated on one side with a green epoxy primer and on the other with red heat-resistant aircraft mastic or adhesive. And it showed no signs of having been burned or overheated.

The mysterious strip of lightweight metal was found at the very start of the long trail of evidence—1,740 meters (5,709 ft) from the threshold of the runway on slab 152—the same slab of concrete where the 4.5-kilogram (9.9-lb) chunk of tire debris was discovered. The metal strip certainly looked as if it belonged to an airplane, but not necessarily Concorde 203. Although runways at Charles de Gaulle Airport and elsewhere are inspected several times daily and cleared of debris as a safety measure, it was possible that metal strip had been dropped

The metal strip that investigators discovered on runway 26R: about 43 centimeters (17 in) long, it punctured Concorde's number 2 tire as it rolled toward takeoff. One large tire fragment then struck fuel tank number 5, in the plane's left wing, while another ripped insulation off a live electrical cable in the number 2 wheel well, resulting in the fire that brought down the plane. Investigators eventually determined that the strip had fallen from the improperly repaired number 3 engine of a Continental Airlines DC-10 that had taken off just before Concorde. (*BEA REPORT*, 61)

between scheduled inspections. Runways at major airports experience a high volume of traffic during the course of a day, and many types of service vehicles come and go along with the aircraft that use the runways to take off and land. Like the other evidence, it was the property of the judicial investigators. "Unfortunately, the Gendarme Transport Aériens, the local police, had put [the evidence] in a safe location," says MacIntosh. "So it was not possible for us to immediately view the damage on the strip or the actual tire parts. But we were provided with photographic evidence that gave us a pretty good idea that we'd had a tire cut, and that tire had most probably come apart in some pretty big pieces that may have done some damage to the underside of the wing." One photograph stood out. "I was quite surprised to see this piece of metal," says MacIntosh. "It didn't look like anything that came from an aircraft to me. Many of my colleagues felt the same way."

6

Too Late

While the runway survey had begun in the fading light of the first day, the arrival of nightfall Tuesday did not interrupt the urgent task of unlocking the vital information contained in the plane's recorders, which were located in Concorde's tail. At 8:40 p.m. on Tuesday, about four hours after the crash, a BEA technician donned an oxygen mask and goggles to protect himself from fumes as he hunted in the charred wreckage for the bright orange boxes containing the CVR and FDR. As soon as he retrieved them, both recorders were immediately placed under seal, and two gendarmes took them to BEA headquarters. The CVR would be opened there, but the FDR would be transported to a laboratory in the south of France. As Tuesday night became Wednesday morning, French technicians at two different locations went to work on opening both boxes, an undertaking imbued with suspense and the shared hope that the magnetic tapes inside had not been damaged. Luck was on their side. The CVR of the voices and sounds heard in the cockpit had been preserved intact, as well as the electronic data on the FDR's tape that provided a moment-by-moment record of the plane's engines and systems on the day of the crash.

The stainless steel case of the Fairchild CVR was badly burned, rendering its serial number illegible. But the CVR's heat-resistant metal box, which is painted bright orange to make it easier to find, had done the job it was meant to do; the four-track tape had not melted.

Midnight came and went as BEA technicians did the vital work of preserving the original recording and also making an enhanced digital copy. In addition to preserving all the sounds in the cockpit and the utterances of the embattled flight crew, the CVR would also allow investigators to listen to radio communication between the flight crew and all airport personnel, including the air traffic controller, as well as any communication between the cockpit flight crew and the six flight attendants in the cabin.

After a duplicate recording was made and enhanced with special software, Air France pilots who knew the members of the flight crew were brought in to verify that the voices heard were indeed those of Christian Marty, Jean Marcot, and Gilles Jardinaud. The Air France pilots were the only people other than those directly involved in the investigation who were allowed to hear the contents of the tape. The emotions of the pilots chosen to listen to their coworkers' disciplined efforts to save the plane and its passengers would not become part of the final BEA report, but it must have been a sobering experience to hear the rapid exchanges among Marty, Marcot, and Jardinaud during their 78-second-long ordeal that began at 4:43:13 p.m., when Gilles Logelin told the crew he saw flames, and lasted until 4:44:31 p.m., not long after First Officer Marcot uttered his last pronouncement: "We're trying for Le Bourget!" In the days that followed, a transcript of the entire recording was made available to the team of investigators.

In addition to the shouted exchange between the three men, the CVR microphone in the cockpit had captured a range of loud, disturbing background noises that had assaulted the pilots' ears. The engine fire alarm had been shut down twice, only to resume ringing and continue unabated to the moment of impact. The cacophony of sounds also included a shrill smoke detector alarm in a bathroom near the cockpit, gongs warning of problems with the plane's pitch, and loud, insistent whoops from the ground avoidance warning system. These and other ambient noises would give Alain Bouillard's teams a more complete picture of the cascade of mechanical failures that the overwhelmed pilots were struggling to manage.

At the same time that the CVR was being opened and analyzed, the Sundstrand-made FDR was being driven to the Bretigny Flight Test Center in the commune of Istres, northwest of Marseille in southern France, some 775 kilometers (482 mi) from the crash site in Gonesse. The government-run center, established in 1944, is home to a number of laboratories and test facilities dedicated to aeronautical research and weapons development. The plane's accident data recorder, or black box, arrived at the Flight Test Center sometime during the night of July 25 or the morning of July 26. It was immediately opened in the presence of two BEA technical investigators, as laboratory technicians set about cleaning the magnetic tape with distilled ethyl alcohol and repairing a small tear in the tape to strengthen it.

Despite their careful ministrations, the condition of the tape was not up to par—the quality of the resulting readout was officially rated as "medium." Because the FDR was an older model, the rate at which sampling or data collection took place was slow compared to newer models. The air speed was being sampled every second, and the FDR collected readings from the engines every second as well, but just one engine at a time—a lag time that would prove problematic for Yann Torres and the others assigned to track the performance of the engines. "The flight recorder wasn't recording parameters on those engines at a rate that we would like to see," says MacIntosh. "It was only once every four seconds." That may not sound consequential, but in fact, the rate of sampling of the engines was later determined to be wholly inadequate when it came to tracking surges in the engines, which can only be identified from fluctuations of very short duration, sometimes of less than a second. As a result, "that caused us a few problems," says Bouillard. The solution was at hand, but it would involve a bit of luck, technical expertise, and the use of a third, newer recorder on the plane.

Fortunately, Concorde was equipped with a more modern device that is known as a quick-access recorder, or QAR, a digital airborne device used by some airlines to store the exact same data recorded on the FDR. Unlike the FDR, which is opened only after an accident, the data stored on the QAR can be accessed at any time and studied to assess a plane's overall performance and operating efficiency. It is not as well protected as the FDR, since it is not meant to survive a crash. In the case of Concorde, the QAR box was crushed and the data storage disc inside was damaged as well. But by sheer chance, one of the QAR's four memory cards, which was visible through a half-torn-off casing, had survived the crash in surprisingly good condition.

Using a method never previously tried, technicians were able to transfer the data from the memory card to a new receiver card. With the help of personnel from the manufacturer, Thomson CSF, a data readout was performed on the card in early August in the presence of representatives of the BEA and also the judicial inquiry team. "Before our examination we had no idea what the QAR contained," says Torres. "Once we succeeded in reading this memory card, we saw by chance that it did contain [the data from] the accident." The successful transfer of data to a new memory card and resulting readout also made it possible to enhance the FDR. "The FDR contained all the data from the flight, but in a state of relatively weak quality," says Torres. "The QAR allowed us to access the same type of information, but in a much more precise way and at a much higher quality." Eventually, the FDR and QAR together would provide a highly detailed history of the second-by-second performance of the plane's operating systems, capturing crucial engine data as well as the fluctuations in air speed, altitude, and other vital aspects of flight.

Recovering the QAR and using the readout to fill in the FDR readout enabled the BEA investigators to examine in detail what Bouillard called "an ensemble of the flight"—its complete operational history. The stage was then set for the investigators to synchronize the CVR and FDR and make connections between sounds in the cockpit in relation to technical and mechanical problems in the plane's engines, whose performance was of particular concern to investigators.

If a loud "bang" was heard on the CVR, for instance, investigators could synchronize that with a related change in pressure readings from the engines associated with a sudden surge, a disruption in the smooth functioning of the engine characterized by a reversal of pressure in the engines' compressors. After having linked the two events on the two different readouts, they could then use that point to synchronize the CVR and the FDR and make other connections between events in the cockpit and multiplying technical problems.

The investigators were quickly given access to all three readouts and the transcripts produced. The CVR essentially validated what air traffic controller Gilles Logelin had already told the investigators during an interview that took place on Tuesday: that the crew was unaware of the flames trailing behind their plane. It quickly became evident that the investigators' initial impressions were valid regarding what the crew did and did not know, observed Bouillard. "The cockpit voice recorder allowed us, very early on, to determine that the crew didn't know about the fire at the rear of the aircraft until warned by the control tower."

After Logelin alerted the pilots to the emergency in progress, they were immediately aware of mechanical dysfunction in the plane's systems, as evidenced by the gauges and warning lights on their instrument panels as well as by the plane's warning alarms, gongs, and bells. "Very quickly they discovered the loss of engine thrust on [engines] one and two," says Bouillard, "and they quickly realized that the plane was flying at a really low speed and that it couldn't ascend."

Less than a week into the investigation, it became clear that the recorders would also verify most of what the investigators had been told by witnesses, which proved useful in terms of corroborating eyewitness accounts and creating a partial narrative of what happened, albeit one that raised as many questions as it answered. The recorders would also continue to provide important insights about how the flight crew responded to the multifaceted emergency, and the immediate problems that arose with the engines. "The flight recorders, notably the cockpit voice recorder, allowed us to understand the actions taken by the crew," says MacIntosh. Because there is always a lag time, however brief, between a malfunction and the human response, the job of establishing precisely *when* mechanical problems arose was paramount—multiple failures were occurring in less than one second—and could only be determined by examining the data on the FDR and QAR. There was no longer any doubt about when and how rapidly the plane's left-side engines had begun to fail, observes MacIntosh. "The flight data recorder allowed us to note, really early in the takeoff, the almost total loss of the thrust from engines one and two."

The CVR held one surprise in the form of a fleeting, unidentified noise that was recorded at 4:43:09 p.m.—just 40 seconds after Captain Marty shoved the throttles forward to maximum power to achieve takeoff. The sound would later be described as a "clean sharp noise," like a pop or a bang, that lasted a mere half second. There was no indication that the crew took notice. But investigators were not willing to dismiss it as inconsequential. Deciphering the source of the noise would later prove critical to understanding the beginning of the cascade of problems that resulted in the fire that consumed the plane.

Even after analyzing the airborne recorders and assembling the debris collected from the runway, the larger picture that Bouillard hoped to conceptually bring into focus was far from clear. "We had all these pieces, but we had trouble making the liaison between these tire pieces, the strip of metal and the pieces of the tank, the kerosene and the traces of soot. All of these remained isolated elements, and we didn't have the link that could unite them. We just couldn't tie it all together." Or, as MacIntosh puts it: "The evidence was not talking to us."

Still, some of the evidence found on the runway—the stain left by the fuel spill, the square fragment from the number 5 fuel tank, and chunks of tread from the left-main landing gear number 2 tire—effectively spoke for themselves. Judging by the oily stain, a large amount of jet fuel had been spilled on the runway before the fire began. The fragment from the number 5 tank on the port side of the plane indicated that the wing tank, filled to capacity for the long flight across the Atlantic Ocean, had ruptured, liberating a steady stream of kerosene, which had been ignited by a spark or possibly intense heat generated by the engines. The large piece of torn tread from one of the plane's massive tires, found lying on the runway 70 meters before the stain created by the kerosene spill, pointed to a tire blowout sometime very early in the series of events. Furthermore, the damaged section of tire looked as if it had been jaggedly slashed by a sharp object. But what could that have been? "There was an initial suspicion that a tire had been involved, and that the tire had some sort of a cut in it that had led to its failure," says MacIntosh.

The tires on Concorde are subject to tremendous pressure upon takeoff. But the Goodyear-made tires on Concorde 203 were not old and worn, nor were they retreads, which Air France stopped using in 1996. They were new and in good condition. Was the mysterious strip of metal found on the runway somehow responsible for the damage to the number 2 tire, resulting in an explosive blowout and unleashing forces that resulted in a rupture in the fuel tank? More investigative work needed to be done to find out why and how the tire had failed and whether there was any connection with previous, well-documented blowouts on other Concordes, especially any that had led to penetration of a fuel tank.

There would also be an intensive effort to try to discover where the unidentified metal strip had come from. In the meantime, diagnosing the problems with engines numbers 1 and 2 remained a high priority. Engine failure had played a major role in the crash. While other teams turned their attention to the burst tire and trying to track down the source of the mysterious strip of bent metal, BEA investigator Yann Torres would head up the effort to find out why the Rolls-Royce/SNECMA Olympus 593 engines had failed to generate the thrust that should have translated into the speed the supersonic plane needed to take off.

The CVR held many clues regarding the performance of the engines, including the sound of the characteristic clicking of the thrust levers in maximum thrust position at 30 seconds after 4:42 p.m.—the moment when Captain Marty applied all of the power at his command to achieve the speed needed for takeoff. There was no doubt that one of the most devastating mechanical malfunctions that had contributed to the crash of Concorde was the loss of power in two of its engines—first number 2 and then later number 1—during and after liftoff. It would be up to the BEA team to find out why the previously reliable engines had suddenly failed. "Looking at the engines was very critical," says MacIntosh, "simply because it [problems with lack of thrust] had been responsible for major activity on the flight deck" in the moments that the aircraft was airborne after liftoff.

There was nothing subtle about Concorde's Rolls-Royce/SNECMA Olympus 593 engines with afterburners, or reheats, which were updated versions of the powerful Bristol-Siddeley Olympus engine that had been fitted to the Vulcan bomber, a high-altitude strategic bomber operated by the Royal Air Force from 1956 to 1984. Developed jointly by Rolls-Royce Corporation and SNECMA Moteurs of France, each engine was capable of producing 38,050 pounds of thrust, or a combined 152,000 pounds of thrust. Concorde was the only commercial airliner in service in 2000 equipped with military-style afterburners that were designed to generate more thrust at key stages of flight, including takeoff. Their raw power and distinctive roar were the source of the window-rattling performance that made Concorde's liftoff unlike that of any other passenger aircraft.

The engines and flame-spitting afterburners had been designed to overcome the lack of lift in the delta wing at subsonic speeds. Any failure of maximum thrust at the crucial moment of liftoff was a disaster in the making. "The Concorde is a plane that requires an enormous amount of thrust in order to move its mass during takeoff," says Bouillard. "Especially for this type of flight with the number of passengers on board."

To reach New York, the plane was taking off at its maximum fuel weight. "The loss of two engines just at the moment of rotation was something that was tragic for the plane," says Bouillard. As a result, "there was a great deal of effort put into examining the engines." Much was at stake in this phase of the investigation. Everything involved with the postmortem on the engines would be carefully overseen by representatives of Rolls-Royce, who assisted in the in-depth postaccident analysis.

If just one of the four engines had failed, the remaining three engines still would have been able to generate enough thrust to allow Concorde F-BTSC to achieve sufficient ground and later air speed to execute a controlled emergency landing at either Charles de Gaulle or Le Bourget. In this respect, Concorde was like any other subsonic plane: It could survive the loss of one engine during takeoff, but not two. "During the takeoff sequence, all planes need their maximum thrust," says Bouillard. "Simply put, the rules of certification stipulate that a plane, no matter what type, Concorde or not, must be able to take off with one broken-down engine, or the failure of a single engine. But there is no mention of taking off with two failed engines."

The possibility of losing two engines was so remote that there was no official procedure in place spelling out how pilots should respond to such an emergency, says MacIntosh. "There was never any consideration that you would lose two engines on the same side simultaneously. That was just an unacceptable event that wasn't planned for."

Then there was the matter of where and how the fatal fire had begun. When it came to the mechanical suspects that might have caused the fire that eventually consumed the plane, the engines and afterburners were number one on the list. All turbojet engines are airborne power plants that burn a highly combustible mixture of fuel and oxygen. The oxygen needed to burn the fuel is ingested through the front intake section of the engine before flowing into the compressor section. From there, the compressed air is mingled with fuel, forming a gas that is ignited in the heart of the engine, the combustion chamber. The hot gases from the engine drive the turbine at the rear of the engine before being expelled through the exhaust nozzle at the rear.

In this respect, the Concorde engines were much like those on other aircraft. But they were very different in terms of where they were positioned. Most engines on commercial aircraft are suspended some distance below the wings in so-called pods. But instead, Concorde's engines were mounted in pairs within rectangular cross-section nacelles attached directly to the underside of the large delta wings, one pair on the left, the other pair on the right.

This made it more likely that an engine malfunction that damaged the blades in either the compressors or turbines could have set off a chain reaction of destruction. If even one blade had shattered and broken free, that could explain the damage to the number 5 fuel tank, which was located inboard and forward in the port wing, with the side-by-side twin engines located at the center rear of the wing. "There's always a concern that some rotating part would be ejected from the engine itself, cause some damage to contiguous parts, and also cause a large fuel leak and perhaps a fire," says MacIntosh. Because the engines were attached to the wings not far behind and adjacent to multiple fuel tanks, they were surrounded by a protective, heat-resistant shield. It would be important to examine the state of the wing, fuel tank number 5, and the turbojet engines at the crash site to see if the shields on engines numbers 1 and 2 had showed signs of overheating or damage.

Engine failure on commercial planes is rare but not unheard of and can result when engines ingest or suck in foreign objects, such as hail, birds, or runway debris. Two well-known engine problems are compressor stall, or a temporary drop in power, and a flame-out in an engine's combustion chamber. These can be corrected by clearing the stall or restarting the engines. In some cases, however, an engine stall can lead to a complete loss of compression and engine surging, a dramatic reversal of the normal front-to-back airflow, and violent expulsion of compressed air out the front intake section of the engine, which creates a loud bang reminiscent of a car backfiring. Locked-in or uncontrolled surging can cause major damage to an engine's working parts, or even a complete meltdown and engine death.

On April 4, 1977, both Pratt and Whitney JT8D turbofan engines on a DC-9-31 operated by Southern Airways surged and failed after the plane flew into the heart of a massive, hail-filled storm system over northern Georgia. After gliding in the powerless plane for some 55 kilometers (34 mi), the pilot, First Officer Lyman Keele Jr., was forced to execute a dead-stick landing on a rural two-lane highway in the community of New Hope, Georgia. He managed a nearly perfect landing, but obstacles on the side of the road clipped the plane's wings and undercarriage, and the jet plowed into a grocery store and gasoline pumps before breaking apart and catching fire on the side of the highway, which was lined with homes and businesses. Nine New Hope residents were killed in the crash, including seven members of one family; 52 people on the plane were killed, including both pilots. It was later determined that the engines had surged to destruction after ingesting large amounts of rainwater and hail.

Engine failure caused by a collision with a flock of Canadian geese resulted in the forced ditching of US Airways Flight 1549 on January 15, 2009 (which was the basis of the 2016 feature film *Sully*, directed by Clint Eastwood and starring Tom Hanks). The Airbus A320-214, with 150 passengers on board, lost all power in both its CFM International turbofan engines when it flew directly into a flock of geese not far from the George Washington Bridge shortly after takeoff from New York City's La Guardia Airport. Captain Chesley "Sully" Sullenberger and First Officer Jeffrey B. Skiles managed to ditch the plane in the Hudson River near midtown Manhattan. All 150 passengers survived and were rescued by nearby boats that came to their aid. The NTSB later found that each of the plane's two engines had sustained damage to their cores as a result of ingesting two birds weighing about 3.6 kilograms (8 lb) each. That resulted in blockage of airflow in the left engine and damage to working parts of the right engine, causing both of them to enter an unrecoverable stall.

With one engine shut down and another surging just 13 seconds after Captain Marty began rotation—raising the nose of the plane—Concorde 203 had encountered significant engine problems before it could begin liftoff and climb-out, which became impossible due to the plane's inability to fly much faster than 200 knots.

Torres and his fellow investigators would start by focusing on why engine number 2 had virtually shut down, producing only 3 percent of its nominal thrust—a rate barely above idle. The problem—a dramatic loss of thrust on two engines reported by the flight crew—began with engine number 2 and later engine number 1, both located on the port side of the plane. Just three seconds after the plane lifted off, the fire alarm began ringing and the flight engineer called for engine number 2 to be shut down. Had engine number 2 overheated and been the source of the deadly fire? Or had the flames originated elsewhere? "At the beginning of the investigation we had, as a hypothesis, a problem with an engine," says Bouillard. "The problems with engines numbers 1 and 2 were very important for our comprehension of the Concorde's accident."

Both the data from the FDR and the frantic verbal exchanges in the cockpit revealed the shockingly rapid drop in power from the left-side engines at precisely the moment when maximum thrust was required to generate the speed needed for takeoff and climb-out. "What was possible to determine, judging from the recorders, was simply that there had been a malfunction in engine number 2 and later engine number 1," says Torres. But *why* had two of the plane's four turbojet engines failed, and why so quickly?

The problem with the engines struck like lightning, manifesting in a single explosive second between 4:43:12 and 4:43:13 p.m., when engines numbers 2 and 1 suffered their first loss of thrust. The drop in power was so sudden and extreme that it caused a sudden, jolting yaw to the left, which was later described as "a violent kick in lateral acceleration." At the same time—4:43:12 p.m.—First Officer Marcot shouted, "Watch out!" as Concorde lurched to the left and Captain Marty reacted by pulling back on the control column, rotating the nose of the plane upward, even though the plane's air speed was only 183 knots—15 knots below the 198 knots prescribed for a successful rotation. Investigators would learn that these dramatic simultaneous developments occurred at the same moment that witnesses reported seeing "an intense luminous phenomenon"—the flash of combustion—accompanied by the loud noise of the engines surging.

All of this took place in the exact same second that the go-lights for both engines number 1 and number 2 blinked off *after* the plane had reached and surpassed V1, the point at which it is impossible to abort a takeoff. Captain Marty had likely reacted instinctively to the sudden, violent yaw to the left in order to avoid a dangerous uncontrolled "lateral excursion" across nearby taxiways, says Concorde pilot Jean-Louis Chatelain. Marty may very well have seen an Air France 747 that had paused on a taxiway about 1,000 meters, or 1 kilometer (3,280 ft, or slightly more than 0.6 mi), distant, waiting to cross runway 26R, and took to the air to avoid a deadly collision. "I don't know if this was taken into account in the decision made by the captain for early rotation," says Chatelain. "But there was a likelihood of an even greater catastrophe on that day with a collision with this 747."

Just a few tenths of a second later, Concorde 203 achieved liftoff even as the malfunctioning portside engines continued to drastically underperform. The plane's four engines were delivering only 50 percent of the thrust needed, with almost all the power being generated by engines numbers 3 and 4 on the starboard side. This sudden lopsided loss of thrust, verified by the recorder data, went a long way toward explaining why the pilots lost control of the plane as it ceased to straddle the centerline of the runway and careened leftward at 183 knots (338 kmh or 210 mph), says Torres. "The fact that engines number 1 and number 2 weren't operating normally allowed us to explain the deviation of the trajectory to the left."

But what had caused the twin portside engines to surge and lose power? Torres considered many possibilities and narrowed them down to three likely scenarios, all of which involved "ingestion," or what the massive engines had sucked in as more power was demanded from them. Engine number 2 had been the first to surge and falter, followed later by the continued surging and eventual failure of engine number 1. "We of course looked at everything involving ingestion, starting with exterior objects that could damage the engine's blades," says Torres. "The second possibility was the ingestion of hot gases inside the engine that would alter its performance and cause it to surge. And the third possibility was liquid ingestion—a direct ingestion of fuel, which could then cause the engine to surge."

Later the engines would be moved to a laboratory to be dismantled. "When we sent the engines to be examined in the lab, it revealed that these engines had suffered a few deformations to their compressors," says Bouillard. Notably, engine number 1 had traces of "gunk" on its compressor blades and deformations—or signs of warping due to exposure to intense heat—that appeared to have resulted when it sucked in both hot gases from the combustion of the kerosene and debris from the blowout, most likely pieces of rubber that had melted and left behind telltale gum deposits on the blades. "During the examination of engine number 2, the deconstruction revealed that the compressor's blades were deformed in a way identical to engine number 1, but without any gum deposits," says Bouillard. That led investigators to conclude that both engines had ingested hot gases generated by the fuel fire, but only number 1 had ingested pieces of tire debris.

Investigators went so far as to order the reconstruction of the badly mangled delta wing, using the damaged pieces of the wing itself, the wing fuel tanks, and the engines. This included an effort to reassemble what was left of the fuel tanks by laying them out on the floor of a large hangar, although very little remained of the badly damaged tank number 5. Special attention was paid to the engine nacelles, or housings, for engines numbers 1 and 2, including the surrounding heat shields, which were undamaged, says Bouillard. There was no indication that the engines had suffered a meltdown or ejected any parts that had punctured the fuel tanks, most of which are located forward of the engines. "There was meticulous investigation of each one of the engine nacelles in a very forensic way to make sure that the engines themselves had not caused additional damage to the aircraft," says Bouillard.

What Torres and his team were looking at were two side-by-side engines that were affected in the same second but in different ways. "The fact that the engines were close together shows that they were likely affected by the same phenomenon or the same cause," Torres says. This "same cause," he decided, was related to "the explosion of the tire," which had sent pieces of rubber flying through the air like projectiles, and which also appeared linked to the damage to fuel tank number 5 and the resulting release of thousands of liters of kerosene.

It was still not at all clear how the fuel streaming from tank number 5 had been ignited. But the preliminary evidence strongly indicated that the engines did not ignite the fire. Rather, the engines gave every indication of being early casualties in the fast-moving sequence of failures that began shortly before rotation and continued through the compromised takeoff, which effectively failed before the aircraft could

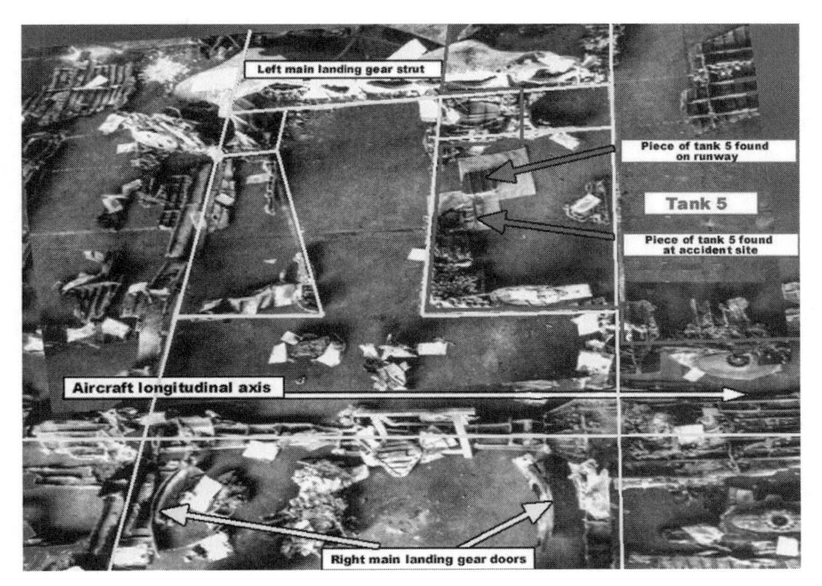

Reconstruction of Concorde F-BTSC by BEA investigators in a hangar at Charles de Gaulle, showing fuel tank number 5, which was struck by a tire fragment when Concorde rolled over a metal strip on runway 26R. (*BEA REPORT*, 80)

continue its normal high-speed climb-out. Instead, Concorde lumbered along, streaming unstoppable flames that spread to other parts of the aircraft, including the left wing. The pilots never had a chance to address the surging in the engines. It would eventually be determined that even if all four engines had been operating normally, the extreme fire damage to the plane's port wing and flight controls would have led to the rapid loss of the aircraft.

Torres and his team had knocked down suspicions that the failed engines were the primary culprits responsible for the crash, although the investigators' attention would later be focused anew on the engines' hot afterburners. By contrast, the blown tire and resulting damage done—the scope of which remained to be determined—had risen to the very top of the list of precipitating events contributing to the crash of Concorde. All of which led investigators to ponder the problem of the unidentified strip of metal found near the remnants of tire number 2.

7

A History of Blowouts

For most of the investigation, Building 153, the nondescript two-story headquarters of the Bureau d'Enquêtes et d'Analyses at Le Bourget Airport, was ground zero for the ongoing inquiry. Although periodic news bulletins would be issued reporting major BEA findings, the details about ongoing research and tests being conducted were strictly confidential. Only investigators were allowed into the restricted area, where they toiled over reams of data and kept long hours. Still, a sense of shared mission and willingness to exchange insights kept everyone sharp and engaged, says Bob MacIntosh. "There was a huge map of the runway pinned to the wall in a BEA headquarters building hallway," he recalls. "Anyone walking by that restricted area was attracted to it to stop and ponder, and was usually joined by colleagues to swap ideas."

BEA investigators had been aggressive in their approach to collecting and analyzing evidence, but cautious when it came to settling on a working hypothesis about the chain of events that had culminated in the destruction of one of the world's most iconic passenger planes. Still, by week two of the investigation, the BEA's seven teams of investigators had decided that the blowout of the number 2 tire had played an important role in transforming a routine takeoff from Charles de Gaulle Airport into an unprecedented disaster.

The mysterious metal strip found on runway 26R had apparently been lying on the pavement during Concorde's sprint toward takeoff.

The fully loaded plane had run over the sharp metal object lying directly in its path, resulting in a slashing cut to tire number 2 and an explosive blowout. That much seemed clear. But where had the strip come from, and why hadn't it been found and removed? Investigators did not know it, but they were about to get a break that would provide the first significant clue as to the strip's origin—a clue that would shed light on what had doomed Concorde but also present them with fresh and troubling difficulties to resolve before they could close the case.

By the end of July, newspapers had moved on to publishing stories about the fallout from the crash and the looming legal battle. "There has never been a crash like Concorde, and there will never be a court fight like it either," Joerg Horny, a Berlin-based lawyer, told reporters for the *Daily Telegraph*. Superlatives had often been invoked to describe the plane itself. But after the crash, they were being used in a different, darker context as journalists predicted that the crash would result in "the biggest compensation battle in aviation accident history," with claims starting at £3 million for every passenger—a grand total of £300 million. "You can tear up the rule book on this one," said Horny. "The claims will be as high as the plane once flew."

Although it was not the BEA's mission to assign blame—that role would be filled by the judicial inquiry—the bureau's investigators knew very well that their findings would be of major importance to all involved, from the relatives of victims to the two major airlines affected and the corporations that made parts for Concorde. The final BEA analysis of the high-profile accident would be endlessly scrutinized and second-guessed. French officials overseeing the inquiry were also expecting progress to be made as quickly as possible. And those directly involved were frustrated at their inability to identify the source of the mystery strip found during the survey of runway 26R that began the day of the crash.

There was no doubt that the bent metal strip was "the root cause of the tire puncture," according to Alain Bouillard, head of the BEA inquiry. This was determined by careful examination that included laying the strip alongside the jagged gash on the largest piece of the number

2 tire to see if the two aligned. "We quickly saw that the imprint of the cut was identical to the shape of the metal strip," says Bouillard. This simple but telling observation was supported by another, says NTSB investigator Bob MacIntosh. "There were black transverse smudges [on the strip] that told us that that piece of metal had been in contact with considerable pressure from a tire. So there was little doubt that the strip and the tire had interacted."

The next logical step was to find out how the offending piece of metal had ended up where it had no business being, and at the worst possible time. "There was a tremendous amount of pressure within the BEA to get to the bottom of where that strip had come from, and we all felt it," says MacIntosh. "I spent more than two weeks at the scene of the investigation, and we still didn't know where it had come from. We hadn't concentrated enough on that yet. It was a big question."

Despite the lingering mystery of the origin of the metal strip, the investigation was moving rather quickly. The evidence gathered on the runway, in conjunction with the recorder data and witness statements, had helped investigators narrow their focus and concentrate on the burst tire and the secondary damage done to the number 5 fuel tank on the inside of the left wing, directly above the burst tire. The fire that had all but consumed the aircraft had been fed by kerosene streaming from at least one large hole in the ruptured tank. The fact that a blowout had occurred on the runway as the plane was picking up speed was significant for many reasons, not the least of which was that other Concordes had been plagued over the years by a string of incidents involving tires that burst or deflated during takeoff.

An extensive review of all related incidents was begun, with a focus on any tire blowouts that had resulted in damage to fuel tanks. The last time the NTSB and BEA had taken a hard look at tires on Concorde was 1979, following a close call involving a Concorde flying out of Dulles International Airport in June of that year. Flight 54, a Paris-bound Air France Concorde charter carrying 81 passengers and a crew of nine, was forced to return to make an emergency landing at Dulles after its two rear left-main landing gear tires burst while the plane was accelerating down the runway. The metal rims on the burst tires were also destroyed.

Metal fragments from the damaged wheel rim had exploded upward and struck the wing, creating a hole 61 centimeters (2 ft) square, according to one witness, and also damaging the number 2 engine and puncturing three fuel tanks—2, 5, and 6—resulting in a small but alarming fuel leak. Several hydraulic lines and electrical wires were severed as well, resulting in partial failure of the plane's hydraulic system. The flight crew was unaware of the burst tire and the damage done to the aircraft until an alert passenger who spotted the hole in the wing notified them. Concorde Captain Jean Doublaits chose to return to Dulles, managing to safely land the plane there by canting it to the right so that the four right-main landing gear tires touched down before the remaining two left-main landing gear tires. Patrick T. Chitwood, an airport operations officer who witnessed the entire episode, later marveled at the smooth touchdown: "The pilot did one heck of a job." Chitwood, who observed the landing from a nearby field, described the mishap as the worst incident involving a Concorde at Dulles since service began there in 1976.

Although the mishap had taken place 21 years before the Concorde crash—and almost a decade before MacIntosh began working for the NTSB in 1988—he remembers many details, including the leak and damage done to the wing. "Back in 1979 at Dulles airport, they had an event that had produced a fuel leak in an aircraft—a Concorde that took off," says MacIntosh, who at that time had recently retired from the U.S. Air Force and was working as an accident investigator for Beech Aircraft Corporation. As a former Air Force pilot and instructor who had trained others pilots to fly the supersonic T-38 Talon, MacIntosh took a keen interest in Concorde and the subsequent investigation into the burst-tire mishap. "It actually had five holes in the underside of the wing, and those holes were traced to a wheel rim—metallic pieces that had broken off of a wheel rim when it had been pressed down on the runway."

In the wake of the blowout, a major effort was made to "get smart" about why the tire failure had caused a potentially deadly fuel leak, according to MacIntosh. Because the plane was an Air France Concorde, France's Direction Générale de l'Aviation Civile (DGAC), the equivalent of the U.S. Federal Aviation Agency (FAA), ordered the BEA to conduct a review of the incident. Investigators looked at the durability of the tires and whether the lining of the Concorde wing was sufficiently protective to guard against penetration in the event of a burst tire. The BEA also considered the possibility of damage to the engines in the wings, as well as the risk of a fire that could be sparked by a fuel leak.

The French safety bureau followed up by issuing several airworthiness directives—official enforced orders notifying a carrier of safety deficiencies that must be corrected. Two of the directives called for the installation of reinforced tires and tire wheels and a tire puncture alarm system. However, the BEA concluded that the risk of fire was "limited" and the possibility of the fuel tanks being penetrated was "sufficiently low"—not enough to require making major structural changes to the wings, both inside and out.

After the Dulles incident, three other burst tire mishaps occurred in the United States over a 20-month period—two more at Dulles and one at Kennedy International Airport in Queens, New York—prompting the NTSB to undertake its own study. The American air safety bureau detailed its concerns in a letter sent to its French counterpart in November 1981. In it, the NTSB voiced its "serious concern" about the four incidents in which Air France Concorde tires had burst during takeoff and the danger of a "potentially catastrophic" fire or explosion that could result.

The NTSB had zeroed in on the threat of fire in the event that an overheated wheel rim were to come into contact with fuel leaking from the large tanks in Concorde's wing. In this type of situation, the NTSB advised, Concorde pilots should refrain from pulling up and stowing the landing gear in order to reduce the possibility that the overheated rims would come in contact with any fuel leaking from the tanks. Improved maintenance, such as checking for proper inflation and routine inspection to detect wear on tire treads, was also recommended.

The safety board also argued that, in cases of tire blowouts, that pilots should be required to return to the takeoff airfield.

The NTSB seriously considered calling for the installation of a heat shield for the fuel tanks and strengthening the exterior of the wing, but it decided against both, says MacIntosh. "Engineering-wise, it would have been very difficult and also very expensive. There was a decision made—which perhaps turned out to be a fateful decision—to reinforce the wheel rim and to not take additional steps to reinforce the lower wing skin and those fuel tanks that were located above the landing gear." After the Concorde crash, it became clear that more could have and should have been done. "The message was that we needed to better protect the wing and we needed to better improve the tires," says MacIntosh. "We didn't get the message."

In the year 2000, the near-disaster at Dulles in 1979 began to look like a wake-up call that yielded some improvements, but not nearly enough. After MacIntosh received word on July 25 that he was assigned to help with the Concorde investigation—and before he got on the plane to Paris—he made a point of picking up the folder containing all the documentation on the Dulles event and personally speaking with the NTSB investigator who oversaw the investigation.

Early on, MacIntosh and the BEA team began conducting their own comprehensive review, which included looking at every burst tire incident over the plane's 24 years of commercial service. It was common knowledge that both Air France and British Airways Concordes had experienced problems with their tires and that the tires were subjected to greater stresses than those on other planes. "The Concorde's tires are unique since they are tires that must handle a load at least as great as that of other planes," says Bouillard, "with the added factor that the Concorde takes off at a speed of about 50 knots more than other [subsonic] planes. So these tires are prone to more incidents."

They soon learned that Concordes operated by British Airways and Air France had been involved in a range of tire failures over the years. No fewer than 57 such incidents had taken place since Concordes began flying in 1976: of those, 47 were either burst or deflated tires, and

10 were instances in which tires lost tread. Thirty of the 57 incidents involved planes operated by Air France, and the remaining 27 were Concordes flown by British Airways. Tellingly, 22 of the 57 tire failures had occurred during takeoff. There were other disturbing similarities to the 2000 crash in Gonesse. One-third, 19 of the 57, tire mishaps were caused by foreign objects on the runway. The 2000 BEA analysis was revealing. Its data showed that problems with burst and deflated tires on Concordes were occurring at a rate of one for every 4,000 flying hours, or about 60 times more frequently than the rate for other long-haul aircraft, such as the Airbus A340.

Still, none of the events had resulted in a major tank rupture or fire. "It was never considered that a tire cut of the magnitude that we saw in the Concorde accident would ever take place," says MacIntosh. "That was simply just not a thinkable event."

Over the course of two decades, the number of tire failures decreased but continued to be a lingering problem, despite the fact that Concorde tires were being replaced after about 35 flights, as opposed to after 92 to 300 flights on subsonic jets. The worst year was 1979, when three planes lost their tire tread and six more suffered from burst or deflated tires. Tire failures persisted with annoying regularity even after the NTSB made its recommendations.

During the 17-year period from 1982 to 1999, there were only six problem-free years in which Concordes did not suffer any blowouts or flat tires. After both airlines stopped using retread tires, the rate of tire failures dropped off considerably in the 1990s, when incidents were occurring roughly every other year at a rate of one to two per year. (Planes experiencing burst tires were reported at a rate of one or two per year in 1990, 1992, 1993, 1995, and 1998. There were no tire incidents in 1991, 1994, 1996, 1997, or 1999. The worst years were 1993 and 1998, when there were two per year.) In 1999, not a single Concorde suffered a blowout. But by early 2000, the chronic problem reared its ugly head again with three burst tire incidents in January, June, and July that preceded the Concorde catastrophe.

The BEA investigators analyzing the fatal 2000 crash were especially interested in cases of burst tires that also resulted in damage to fuel tanks. It was, after all, not the tire blowout in itself that had proven fatal to Flight 4590. Rather, it was the combination of the damage to the wing tank and resulting massive fuel leak that made the 2000 incident different and deadly. By combing through records kept by the two operating airlines and also by the maker of the tire, Goodyear Tire and Rubber Company, investigators were able to determine that between June 1979 and July 1993, just six tire failures had resulted in pieces of the burst tires or other debris causing structural damage to fuel tanks or penetrating them.

On October 25, 1993, a British Airways Concorde taking off from London's Heathrow Airport suffered a brake lock, which caused a main landing gear tire to burst, damaging the water deflector. A piece of the deflector holed the number 1 wing tank, located above the wheels on the port side of the plane.

In most cases, with the notable exception of the 1979 event, the damage to fuel tanks was minor. Still, buried in the reports were details that shone a light on the potential for a wide range of secondary damage. In four of the six cases, the damage to tanks was not caused by pieces of the burst tire but by other bits and pieces of the plane that had broken loose or were shattered and transformed into shrapnel-like flying projectiles. Those included a bolt that punctured a fuel tank, pieces of a wheel rim, pieces of the door to the landing gear, and fiberglass fragments from a shattered water deflector that caused minor damage to a fuel tank.

The investigators also uncovered another similarity between past tire blowouts and the burst tire episode that caused the 2000 crash. Five of the six instances that resulted in damaged fuel tanks had occurred when the planes were taking off; only one tire had burst during a landing. Unlike on Flight 4590, none of these tire failures had resulted in loss of life or serious damage to the aircraft. Instead the planes had continued their takeoffs and circled back to land safely, although it's very likely that these pilots—like Captain Doublaits in 1979—were not at first aware that one of the main landing gear tires had burst. Only one pilot had chosen to abort the takeoff after a tire burst. The investigators needed to find out why the tire burst on Concorde 203 had caused a major rupture of the fuel tank—a problem that would require a much more aggressive approach to analyzing the chain of events unleashed by the blowout on runway 26R.

In the meantime, investigators had in their possession the metal strip that had caused the tire to explode, and they could not rest until they tracked it down to its source. By week two of the inquiry, they were certain that it was not a damaged part from Concorde 203 itself. But if the strip did not belong to Concorde, where had it come from? It was a simple question that proved frustratingly difficult to answer. The other troubling question concerned why a 43-centimeter (17-in) piece of metal debris had not been seen and removed before Concorde went thundering down runway 26R.

Most airports, especially high-traffic hubs such as Charles de Gaulle, expend considerable effort clearing away objects dropped on runways by service vehicles and sometimes, although more rarely, by the planes themselves. "We find that a certain number of objects are lost by planes on the runways during taxi," says Alain Bouillard, whose job was to oversee the wide-ranging BEA inquiry. "It's rare enough. But it's not trivial. Put simply, it's one of the things that we seek to identify with inspections or other means, including detection by radar, because we know by experience that these dropped pieces can lead to catastrophe."

Charles de Gaulle had an adequate but not especially aggressive cleaning program that called for three daily inspections of all four runways at flexible times in the early morning, at midday, and during the late afternoon, sometimes before sunset. This hewed to international standards that called for a minimum of thrice-daily runway inspections at an aerodrome the size of Charles de Gaulle but was not as diligent as other airports, including Heathrow Airport in London, which in 2000 inspected its runways five times a day. "The international standard called for three-times-a-day inspections of the runway surface all the way from the parking lot out to the runway itself and to the far ends to make sure we didn't have foreign objects lying around, ready to damage an aircraft on the takeoff or even in the taxi area," says MacIntosh.

On July 25, the day of the accident, the morning inspection had been carried out at 4:30 a.m., followed by a partial inspection at 2:30 p.m., due to the report of a bird strike. A full inspection scheduled for 3 p.m. was postponed because of the fire drills that took place on runways 26R and 26L between 2:35 and 3:10 p.m. As a result, runway 26R had not been fully inspected for more than 12 hours prior to the Concorde takeoff at 4:42 p.m.

The fact that a piece as large as the bent metal strip had not been noticed was surprising in itself, says MacIntosh. "During the days of the reciprocating engines, there were frequently parts falling off engines, including exhaust stacks and various pieces of debris that would get out on the runway and would be obstacles for the next aircraft. With the advent of the jet age, most of those things got better. Certainly, we got less and less debris. By the year 2000, things had improved to the degree that most of the debris that was picked up off runways would be little things like static wicks [wires screwed to airframes of planes to discharge static electricity] and little pieces of rubber, or perhaps an occasional bottle cap left lying around during servicing of an aircraft. Normally, you didn't find any large pieces of debris on a runway."

To solve the riddle of how the strip ended up on the runway, Bouillard's air safety detectives relied on a process of elimination. They began by agreeing that the strip did not belong to any service cart or other ground vehicle but showed every indication of having been riveted onto an aircraft. First, the strip was made of titanium, a type of metal used on aircraft because it has the strength of steel but is lightweight. Second, the 12 rivet holes along the length of the strip proved that it had belonged to an aircraft, as did the type of mastic, or adhesive, that coated one side, likely applied to help hold it in place. The investigators paid careful attention to the possibility that it had come from Concorde 203 itself, but that was quickly ruled out by checking maintenance records, which revealed no such strip in the aircraft. "We knew early on that it was a strip of metal *not* belonging to the Concorde and that it didn't belong to any terrestrial vehicles," says Bouillard. "So we began looking at the planes that had taken off before the Concorde."

Although the runway had not been fully inspected for 12 hours, the investigators suspected that the unidentified strip had not been lying there for very long. Thus, they started to look at the planes that had been on the same runway as Concorde on the afternoon of July 25. First they focused on the planes that had taken off on runway 26R prior to Concorde, and the same investigator who had found the orphan strip began his hunt for its parent. "He had been working on this from the get-go," says MacIntosh. The investigator looked at the two aircraft that had taken off immediately before Concorde: a Boeing 747 jumbo jet, which was the last plane to depart before Flight 4590; and a McDonnell Douglas DC-10 wide-body, which had taken off just prior to the 747. Because the 747 was owned by Air France, investigators were able to quickly determine that the plane was not missing any parts or metal strips used in a repair. "Originally, we hadn't had any suspicion towards Continental's DC-10," says Bouillard. "We simply knew that it had taken off before the Concorde." But that quickly changed.

The critical gaze of the crash investigators shifted toward the DC-10, operated by Continental Airlines, which had taken off just before the Air France 747 on a flight to Newark, New Jersey. It took two weeks of searching airport records to find the plane's N number, or registration number. That enabled the investigator conducting the search to learn that the same DC-10 was scheduled to return to Charles de Gaulle Airport on August 30. The BEA investigator, a civil engineer, had a ramp access pass and clearance arranged with airport security, so he was able to conduct a quick inspection of planes arriving or departing. Once he knew that the DC-10 was making a stopover in Roissy, "the investigator went to see this plane," says Bouillard, "and he observed that there was a piece that was missing on the rear hood of one engine."

"He got a quick glimpse," says MacIntosh. "But he had enough of a glimpse to say that he thought there was something wrong." That something was a hint of red mastic glue material around the number 3 engine that seemed to match the red mastic on the strip. "It must have been a very special moment for him," says MacIntosh, "because he'd been spending a lot of time and effort looking at aircraft and trying to figure out where this piece of metal might have come from. He was elated, and he rushed back to the office to let everyone know about this particular aircraft operated by Continental Airlines. He thought he had the golden nugget, and indeed he did. He called us almost immediately and said, 'I think I found the aircraft, and indeed it's an aircraft part. It's off a DC-10, and it's a strip around the engine.'"

MacIntosh received the phone call about the discovery on a Thursday, soon after his arrival at NTSB headquarters in Washington, D.C. "I came home to NTSB after a long stay in Paris in August—taking a break from the on-scene investigation. I was sitting in my office when I received a call from the BEA investigator. He gave me details of his investigation to date and said he spotted the area on the nacelle where he thought the strip should have been located. He wanted our help to locate it and hold it for inspection."

The news from Paris kicked the investigation into high gear, with MacIntosh in charge of tracking down the whereabouts of the DC-10 and persuading Continental Airlines to let the French investigators inspect the plane. "It was going to be my job to try and put things together to get the eyes on this aircraft, wherever it was. It was going to be a challenge. But on the other hand, that's what we'd all been looking for. So we felt good. But we also felt concerned because we were going to have to unfold a whole new chapter in the investigation."

At that point, he says, "it really became a challenge to me. Before, I had a duty as an accredited representative, but now I was shifting to an investigator-in-charge mode in order to find the specific point where this strip had come from, and also find out about the maintenance history that allowed this part to escape from the aircraft and cause the damage that we had all seen." He and French investigators still in Paris began making the necessary calls. "We were all glued to the phone, trying to figure out when we could look at the aircraft, because by this time it had left Charles de Gaulle and was flying around in North America."

MacIntosh contacted key Continental personnel, who were immediately willing to help the investigators. "I called Toby Carroll, the Continental Airlines director of safety, and we located the airplane." No time was lost thereafter; the aircraft was due to arrive Saturday at Houston George Bush Intercontinental Airport, commonly referred to as Houston Intercontinental Airport (IAH). "We agreed to meet," MacIntosh says, "and I rounded up the FAA investigator, Bud Donner, for early Saturday travel to Houston. The BEA guy came from Paris to Washington Friday night and joined us for Saturday travel. So that's how we all got together at IAH—waiting for that aircraft to block into its parking spot on the ramp."

Continental's safety manager had been forthcoming about recent repairs made to the DC-10 that had taken off that same afternoon, July 25, from runway 26R, says MacIntosh. "After a quick handshake, we all realized that we had a pretty serious duty here, and there might be some real repercussions coming from what we were about to do." The group went down onto the ramp, where they could observe the aircraft as it rolled in. "Of course, all of our eyes were trained on the suspect area on the engine. My French counterpart was there also." MacIntosh and his French colleague approached the airplane, which was filled with passengers preparing to disembark. "We walked out on the ramp while the passengers were still deplaning. Wow, what a surprise! There was the little gap visible in the nacelle cowling doorframe. And upon opening the cowling, there was the place where the strip was missing. As soon as we knelt down, we saw this red mastic material and recognized that 'Yes, this is the place where that very damaging piece of metal had probably escaped from.'"

Once the passengers were off the aircraft, a more leisurely inspection was in order. "We had the mechanics open up the cowling and expose the area where the strip was missing," says MacIntosh. The purpose of the wear strips was to make the cowling fit more tightly and at the same time to allow the cowling, which is the cover of the engine, to move slightly and be flexible so that it could grow or shrink in response to changing temperatures. "That's why it's called a wear strip," says MacIntosh, "because it would eventually wear away."

The investigators knew what the strip looked like and had brought along photos of the strip, which was punctured by a series of irregularly drilled holes. "Consequently, we compared some of the holes on the piece on the engine of the Continental aircraft with the holes that were in the photographs that we had of the metal strip. And there was no doubt in our minds that the strip had come from this aircraft. There was the red mastic. There was a bit of soot in certain holes, and so on. There was no doubt."

MacIntosh and the BEA investigator were elated. "There was a considerable amount of relief among those of us from the NTSB and from BEA. We'd found the culprit." But the others present did not share their sense of satisfaction at having solved the mystery. "Immediately it became an issue for Continental Airlines and the FAA as to how such a thing could have taken place." The representatives from Continental Airlines, including the director of maintenance and the chief mechanic, were all "very apprehensive," says MacIntosh. "Because they knew that something had gone wrong in their maintenance procedure that had allowed this strip to come loose. So there were some very different emotions there at that moment."

All of the BEA investigators' hard work and persistence had paid off. "We knew we had the right part," MacIntosh recalls. "It was very important to all of us, and we immediately made phone contact with Monsieur Alain Bouillard, the investigator in charge, to tell him that it was very conclusive that we had found the source of the strip," says MacIntosh. The discovery of where it came from was "a turning point," says MacIntosh, "because it further reinforced the interrelationship of the strip and the burst tire and everything that happened afterward. Something had struck the bottom side of the tank and caused this massive fuel leak, which was further supported by the fact that a piece of the tank was found along with the piece of rubber tire. That caused this huge amount of fuel to come out of the underside of the wing. All these things were connected."

Because the metal strip found on the runway was in the possession of the French gendarmerie, it was not possible to bring the strip itself to Houston and lay it over the engine to see if the holes in the strip lined up with the rivet holes on the engine. "Just to make sure that we had the right part, we made a map of the holes by taking a pencil and paper and rubbing the pencil over the spot where the holes were and the width and so on, to make an impression, so that the two could be compared and lead to a further determination of conclusive evidence that we had the right spot on the aircraft," MacIntosh says. "We wanted to make sure that we had the right aircraft, that indeed the strip matched, so we also made some scrapings of that red material. And we made some scrapings of the paint that had been applied at the same time. So we had paint materials. We had scrapings of the glue, and those were carried back to France, where they were compared in a laboratory to the materials that were in evidence on the cut tire."

Even before Continental conducted its own review of why the repair had failed, MacIntosh saw several problems with the basic workmanship. The rivet holes on the engine didn't match up with the holes that the mechanic drilled into the strip. Also, the repair strip was too long, "so it kind of stuck up" from the forging on the engine. "There were things going wrong there from the get-go," says MacIntosh. In addition, the wrong metal had been used to manufacture the strip. "It was supposed to be stainless steel, and it turned out that the mechanic drew some stock from titanium, which has different strength properties," says MacIntosh. "Also, there were a lot of rivet holes that were no longer used, and according to the maintenance instructions, should have been filled in, but those weren't. So there was incomplete and sloppy maintenance to put the strip back on." When it was apparent the strip didn't fit quite right, the mastic was applied to it to make sure that it was glued on. "Obviously that was unsuccessful, and the strip eventually worked loose."

No one inspected the repair, made on July 9 in Houston, because it was a routine, noncritical part that didn't affect the safe operation of the DC-10. As long as the repair was made, that was good enough. "You can see the consequences of that attitude," says MacIntosh. "On the other hand, there was no intent to violate a regulation. It was just that the job wasn't done to a professional level that you would expect from a first-line airline in particular." Later, investigators would find out that the metal strip had been replaced more than once, says Bouillard. The previous replacement of the wear strip took place in June 2000 on a stopover in Tel Aviv, Israel. "That was the reason why this mechanic had changed the material and cut a strip out of titanium instead of stainless steel, which was the recommended material for this piece of equipment. So the use of titanium was done in good conscience by the mechanic. It seemed to him that a material such as titanium would be more resistant and would prevent another loss of this strip." Ironically, the faulty repair did not pose a problem for the DC-10 itself. "The plane had flown for many days without the metal wear strip, but that had caused no danger or affected the safety of the flights," observes Bouillard.

Still, if it had been found in time, picked up, and removed, the crash would not have happened, which raised another, seemingly minor matter—the fact that the midday runway inspection did not take place because of fire drills. A great deal had been accomplished in Houston, but more needed to be done. Continental Airlines would have to review its maintenance records and policies. And the investigators wanted to finalize their investigation into the wear strip.

Rigorous accident recreation tests would be performed in Akron, Ohio, and Toulouse, France, to establish that the wear strip from the DC-10 was capable of inflicting the same type of damage under similar circumstances. The BEA fabricated a metal strip from titanium and gave it the same shape as the strip found on the runway. "The next step was to go to Goodyear in the United States, in the presence of a representative of the NTSB, to do rolling tests with tires identical to those on the Concorde," says Bouillard. The tests were carried out at the Goodyear technical center in Akron using a tire test truck and trolley loaded with metal weights of 907 kilograms (2,000 lb) and up. The beams were distributed so that each of the tires was bearing a weight of 25 tons each—equivalent to the weight on Concorde's main landing gear tires.

Next the truck and trolley were driven over the bent titanium strip, which had been stood on edge on a concrete surface. On the first try, the tire simply flattened the strip. But the second time, the strip's sharp edge tore into the tire, leaving behind a slashing cut whose imprint matched the curved shape of the strip. The damage to the tire was extensive, and included the shoulder and sidewalls. The cut went as deep as the tire beads—the rubber-encased steel wires that keep the tire affixed to the rim. The technicians conducting the test then examined the debris and quickly concluded that the size and weight of the pieces of new Goodyear tire used in the test were very similar to the tire debris found on the runway on the day of the crash.

For further verification—and because the Goodyear test truck could not duplicate the speed of Concorde—other tests were carried out at the Aeronautical Test Center in Toulouse. Technicians used metallic strips identical to the wear strip that were mounted on metal plates and cables that were run between a metal drum and a tire rotating at high speeds, approximating the speed of the plane prior to takeoff. The first tire tested immediately burst, as did the second and subsequent tires. The ejected chunks of the burst tires from both Akron and Toulouse were then sent to another French laboratory, the Rubber and Plastics Research and Test Laboratory, for comparison to the rupture patterns on the remains of Concorde tire number 2—with positive results.

In addition, an analysis of the primer paint from the engine of the DC-10 showed those paint samples were similar to the primer paint on the dropped wear strip; the red mastic sampled was also analyzed and found to be a silicon mastic of the same type found on the wear strip. The investigators were satisfied that the evidence and tests together established that Concorde's number 2 left-main landing gear tire had burst when the plane rolled over the wear strip that had fallen off the DC-10 that had taken off on the same runway less than five minutes before Flight 4590. On August 4, the BEA had issued a news bulletin confirming that the metal strip that resulted in the burst tire did not belong to Concorde—a source of relief to Air France management, who no doubt would be eager to shift at least part of the blame for the crash to Continental Airlines.

It was not long, however, before Air France's own maintenance practices came under scrutiny. The results would be startling. While investigators were busy tracking down the source of the dropped wear strip, a massive effort was underway to reassemble the various parts of the plane using the wreckage collected at the crash site. At the same time, other BEA employees were systematically reviewing maintenance records for Concorde F-BTSC. What they found was disconcerting and would require investigators to take a fresh look at why the plane had gone off course and veered to the left, striking a light on the left side of the runway just before it lifted off. While technicians were reassembling the left-main landing gear system, it was discovered that there was a part missing from the center coupling that connects the landing gear wheel assembly, or bogie, to the vertical strut that attaches the landing gear to the fuselage.

The missing part was a spacer, or a metal sleeve inside the coupling. The absence of the spacer was not a small matter, given the circumstances of the crash and the spacer's purpose, which is to maintain the proper position of two shear rings located at each end of the spacer. The spacer maintains proper alignment of the landing gear and prevents lateral or sideways movement of the wheels, which could result in wheel wobble that might make the plane difficult to control and contribute to tire overheating and damage.

A check of maintenance records showed that the spacer had not been reinstalled during a major repair that had taken place between July 17 and 21. The entire left-main landing gear bogie beam or leg had been removed and replaced with a new leg in order to correct a minor problem related to the aircraft's underinflated-tire detection system. It was the first time that a bogie had been replaced on a Concorde operated by Air France, and the results had been far from satisfactory.

When the BEA reported the news about the missing spacer, reporters began asking the obvious questions. Had it caused the aircraft's left-main wheels to wobble and drag the plane to the left? Was

Concorde

COMPONENT LOCATION HANDBOOK

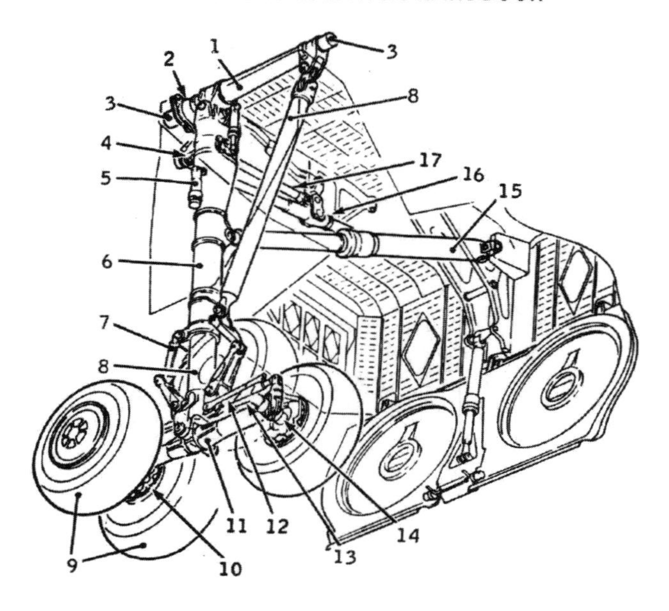

1	CROSS BEAM	10	BRAKES
2	UPPER BRACE TRUSS ROD	11	BOGIE-BEAM
3	TRUNNION	12	BRAKE TORQUE LINK
4	LOWER BRACE TRUSS ROD	13	PITCH DAMPER
5	HYD. SEQUENCE VALVE	14	AXLE
6	LEG	15	TELESCOPIC BRACE STRUT
7	TORQUE LINKS	16	ACTUATING CYLINDER
8	SHOCK ABSORBER	17	ACTUATING ROD
9	WHEELS	18	DRAG STRUT

MAIN LANDING GEAR

Concorde's landing gear. Investigators discovered that, during a repair to Concorde F-BTSC's left-main landing gear that took place between July 17 and 21, 2000, a spacer was not reinstalled in the center coupling that connects the landing gear wheel assembly, or bogie (11), to the vertical strut that attaches the landing gear to the jet's fuselage. This error became a focus of the investigation before it was determined that the spacer's absence had not contributed to the crash. (LENEY AND MACDONALD, *AÉROSPATIALE / BAC CONCORDE*) it possible that the wheels were already overheated when the plane ran over the metal strip, exacerbating the force of the blowout? Both possibilities seemed likely, given the difficulty that Captain Marty had had in keeping Concorde under control after the left-main landing gear tire exploded.

The omission of the spacer was a serious mistake that could not be glossed over. The discovery sent investigators scurrying off to evaluate what had happened and all of the possible problems that could have resulted. Among other things, it was learned that the repair was carried out in the course of two 12-hour shifts, which complied with regulations. However, it was also discovered that one of the prescribed tools needed to do the job properly was not used. All the required postrepair tests and checks were made, but no problems were detected, and Concorde 203 went back into service without the spacer.

To evaluate the effects of the oversight, an examination of the left-main landing gear assembly was undertaken by BEA investigators at the testing center in Toulouse, while a concurrent study was made of all four previous flights of Concorde 203 between July 20 and July 25. The examination showed that one of the shear rings, which was no longer securely held in place, had moved incrementally over the course of four days and four flights. Investigators also studied the ground trajectories of the plane during those four flights. There was no indication the plane had shown any signs of wheel wobble or controllability problems.

A further check of the FDR data revealed that brake temperatures were normal for both the left and right bogies while the plane was taxiing to its holding position on runway 26R—an indication that no abnormal braking was needed to counteract a leftward drag that would have resulted if left wheels were out of alignment. The FDR data also showed that Captain Marty did not need to use the right rudder to stay on course *before* the tire burst. And no tire skid marks were found on the runway before the tire burst—another indication that the wheels were not pulling to the left. In addition, investigators reviewed the documented track of the aircraft during the first two-thirds of the takeoff, before the plane ran over the wear strip. But a review of the data on the FDR during this time showed no unusual lateral acceleration or odd sideways movement, sometimes called crabbing, that would have resulted from instability in the left-main landing gear. In addition, the CVR transcript revealed no conversation in the cockpit about anything unusual in the handling of the plane prior to the blowout.

In the end, the BEA investigators found considerable fault with the sloppy maintenance that had led to the omission of the spacer, but they concluded that the missing part had not been a contributing factor in the accident. Among other things, the research conducted by the BEA demonstrated that even the maximum possible force that could have been generated by the bogie's pulling to the left due to the missing spacer was trivial compared to the forces being generated by the starboard-side engines. It was a finding that was supported by the best available evidence but which would later be hotly disputed in media reports and by critics of Air France and the BEA investigation.

•

The Concorde disaster had shaken public confidence and cast a dark shadow on the reliability of the two aging fleets of Concordes operated by Air France and British Airways. Air France had voluntarily suspended all Concorde flights after the accident, while British Airways Concordes continued to make twice-daily flights between London and New York. But that was about to change. On August 15, not long after the BEA identified the burst tire as the precipitating cause of the crash in Gonesse, the AAIB and the Civil Aviation Authority notified British Airways that its certificate of airworthiness for Concorde would be revoked. The airline responded by immediately grounding its fleet of Concordes. Last-minute arrangements had to be made for passengers when the airline cancelled a 10:30 a.m. flight from London to New York as the plane stood on the runway at Heathrow Airport preparing to take off.

Legal battles over what and who was to blame for the Concorde crash were also proceeding apace. On September 27, Air France and the airline's insurer, Réunion Aérienne Consortium, announced that they had filed a complaint against Continental Airlines in a civil court in Pontoise, just north of Paris. "A piece from a Continental Airlines plane could have caused the accident," a spokeswoman for Air France said. The accusations were the early shots fired in what would become an all-out war—what Bob MacIntosh called "a lawyer frenzy." For its part, Continental issued a cursory statement to the effect that it was aware of the pending legal action but insisted that "at this stage of the investigation, there is no conclusive evidence that Continental Airlines was involved in the crash."

Privately, investigators were confident that they had done more than establish probable cause regarding what had led to the first critical mishap in the chain of events that began with a burst tire and ended with the catastrophic destruction of a plane filled to capacity with carefree holiday travelers. Using an approach that was both systematic and exhaustive, they had conclusively identified the wear strip found on the runway had been improperly affixed to the engine of a DC-10 in a Continental Airlines maintenance hangar in Houston on July 9. It had remained in place for more than two weeks before it came loose and fell from the same plane on Charles de Gaulle's runway 26R on July 25, just minutes before the controller gave Flight 4590 the go-ahead to enter the runway and commence takeoff.

When that determination was made, other odd and troubling aspects of the crash emerged to serve as reminders of how happenstance and entropy can conspire to wreak havoc. If the metal wear strip had been lying flat instead of upright on its edge, the speeding plane would most likely have passed cleanly over it. And even if there had been no fire drill that day and the 2 p.m. runway inspection had been carried out on time, it would have made no difference. The strip had been lying on the runway for less than five minutes when the number 2 left main landing gear tire ran over it as Captain Marty accelerated down the runway at about 170 knots (315 kmh or 195 mph).

The investigators had made significant progress in their quest to reconstruct the series of events that had led to the devastating loss of so many lives and the destruction of *l'oiseau blanc*. But their task was far from complete. Many other Concordes had suffered similar tire failures that had not caused the sort of damage that resulted in a massive fuel leak that, once ignited, had burned out of control, releasing hot gases that overwhelmed the portside engines and crippled a plane built to fly faster than the speed of sound.

"Our initial look at the Concorde scenario presented a lot of questions about how such a major amount of fuel got liberated from the wing of the aircraft," says MacIntosh. "The metal strip and the tire didn't explain everything that had taken place. It simply presented more questions to us as to how that piece of tire could damage the wing so extensively that it would liberate that much fuel and cause such a conflagration in the back of the wing." Bouillard summed it up this way: "By drawing a link between the metal strip found on the runway and the pieces of tire, we had simply accomplished the first step. We had two things left to understand, the first being how tank number 5 had been perforated. And the second thing was how the fire could have started."

8

An Invisible Shock Wave

Like homicide detectives pondering anew the scene of a crime, the air accident investigators went back to runway 26R and began reviewing evidence and looking at their notes. "We had a big fuel stain on the runway prior to the ignition, then all of this smoke and fumes and so on, which had deposited soot on the runway," says Bob MacIntosh. They also had a large, square piece of the number 5 fuel tank, built into the wing over the left-main landing gear tire that had burst. "We knew tank number 5 had been perforated for the good and simple reason that we had found a piece of this reservoir on the runway," says Alain Bouillard. The square, ribbed piece was 32 by 32 centimeters (12.6 by 12.6 in)—big enough to produce a hole that could release a significant quantity of fuel. More tellingly, the square piece of the tank was remarkably clean and undamaged, without any deformations. "There were no signs of shock on this piece," says Bouillard.

How was that possible? If a 4.5-kilogram piece of the tire had flown upward at an estimated speed of more than 140 meters per second (460 ft/sec, or 313 mph), with enough force to penetrate the wing and hole the aluminum tank, any fragment created by the impact would surely have been dented or even crushed. Instead, the square fragment looked as if it had been cleanly cut away from the bottom of the tank. Oddly enough, the fragment gave every indication that the force that caused the rupture had come not from outside but from *inside* the tank. "Our subsequent analysis indicated that this rupture had taken place because of a thrust or pressure coming from the interior toward the exterior," says Bouillard. But this was puzzling, as there was nothing inside the tank but fuel and air.

To solve the riddle, investigators came up with a theory involving a shock wave acting on the inside of the tank, creating an active hydrodynamic force that ruptured the tank. In other words, a powerful shock wave moving through the compressed fuel inside the tank had exerted enough pressure concentrated in one place in the tank, which was filled to capacity, to create the square hole through which the fuel had gushed. The generation of the shock wave was directly linked to the molecular properties of fuel, which, like other liquids, cannot be compressed. "Basically, when you press on a surface that's filled with liquid in one spot, it produces a reaction in other places because liquid is not compressible," MacIntosh explains. Since the tank was full, the energy created by the shock wave could not be dissipated. "If the tank had been three-quarters full, the convection of fuel would not have exerted this pressure on the body of the tank," says Bouillard.

The shock wave theory made sense to investigators, given the evidence at hand. They also considered damage that had been done to military planes under fire in wartime. Military aircraft designers had learned the hard way that shock waves from high-velocity, low-mass bullets can cause fuel tanks in military aircraft to fail *outward* from indirect internal pressure rather than *inward* from the direct external force caused by a bullet. A flying fragment from a burst tire is heavier than a bullet, and it does not travel as fast. But the Concorde investigators suspected that a similar failure effect could have been caused by the higher-mass, lower-velocity tire fragment that produced a high-pressure shock wave within the tank, ultimately rupturing it outward, as suggested by the recovered tank fragment.

Investigators were well aware that the tank had been slightly overfilled on the day of the crash to maximize the quantity of fuel for the long flight. "In order for the Concorde to reach the other side of the Atlantic, its fuel tanks must be filled to the maximum," says BEA investigator Yann Torres. "That means that there's very little free space in the tank for air." As Concorde 203 picked up speed on the runway—hitting speeds of 100 knots, then 150 and 180 knots—the kerosene in the tank was forced to the rear, effectively displacing any residual air in the tank toward the front. This meant that the rear of the tank, where the shock wave was directed, was also the area most susceptible to hydrodynamic shock.

The theory of hydrodynamic action allowed the BEA investigators to understand how the shock of a large piece of tire striking the exterior of the tank had exerted pressure on the tank interior at a secondary point removed from the site of impact. When this possibility was considered, everything seemed clear and the evidence fell into place. "It was a revelation to all of us, and a very plausible explanation as to how the major fuel liberation took place," says Torres.

In addition to the 4.5-kilogram piece of tire that went flying through the air, smaller pieces of the tire and fiberglass fragments of the shattered water deflector could have acted like shrapnel striking the wing. All these phenomena added up, Torres says, "so that we're convinced that was indeed the puncture, the piece of tire, and maybe another projectile as well, that combined to produce the perforation of the tank." The investigators were optimistic they had the answer, but they were also cautious, aware that they were breaking new ground. "In fact, we had never seen it in a commercial aircraft before," says MacIntosh. "It's not a common phenomenon," says Bouillard. "But we were convinced that there was a very high probability that this phenomenon had occurred and was the reason for the fuel leak."

The next step was to put their theory to the test at the Centre d'Essais Aéronautiques (CEAT, Aeronautical Test Center) in Toulouse. Investigators decided to employ an effective albeit somewhat crude device sometimes called a chicken gun—a large-diameter cannon that uses compressed air to fire chicken carcasses at windshields and other aircraft components, such as engines, to simulate a bird strike. Instead of a dead chicken, the investigators loaded the cannon with chunks of tire of about the same weight as the largest, 4.5-kilogram piece found on the runway after the crash. "We conducted two main tests," says Bouillard. "Test number one went down in Toulouse with a cannon usually used to certify plane engines." A full-scale model of the number 5 tank was constructed and filled with a liquid with the same density as kerosene and equipped with sensors that would record the effect of impacts on the tank lining.

"We loaded the cannon with a piece of tire in the same weight range as those found on the runway—4.5 to 4.8 kilograms [9.9 to 10.5 lb]. And we shot the pieces of tire onto a tank of the same type as the one on the Concorde." While the model tank did not actually burst, it did show visual evidence of outward deformation consistent with the shock wave theory. "We were not able to reproduce what had happened in Roissy because of the limits of the installation," says Bouillard. "We could only shoot at a speed of 106 meters [348 ft] per second, which was the maximum speed of the cannon," whereas calculations showed that when the tire burst, the pieces of tire had reached speeds of 140 meters per second (460 ft/sec or 314 mph). "So we think today that if we had been able to take a shot at this higher speed, we would probably have reproduced the phenomenon of perforating the number 5 on the Concorde during takeoff," Bouillard says. Nonetheless, the sensors inside the tanks clearly indicated that investigators were on the right track. "There was an area that was comparable to the hole found on tank number 5, with significant force concentrated in this area."

Investigators hoped they would get more precise results using computerized modeling, which would allow them to program in data that exactly replicated the dynamics of the tire burst. Computer programmers were put to work modeling simulated pieces of tire that would travel at 140 meters per second and slam into a simulated tank. "We took a virtual shot with a [simulated] piece of tire equivalent to that which was found on the runway at Roissy," says Bouillard. Investigators looked on with a mixture of anticipation and anxiety that turned to delight as the animated screen display showed a shock wave forming on the lower surface of the tank with enough force to cause the simulated tank to abruptly rupture. "It was perfect—just fantastic!" says Torres, summing up the investigators' collective response to the test results.

While testing supported the shock wave explanation of the tank rupture that fueled the blaze, this was not the entire answer. One badly burned and visibly damaged section from the underside of the number 5 tank was also recovered from the crash site in Gonesse. The section of the hull had been punctured, leaving behind a hole 10 millimeters wide by 40 millimeters long (0.3 by 1.5 in). This led the investigators to conclude that the tank had been holed as a result of a direct impact by another piece of debris. An examination of the small rectangular hole in the burned section recovered at the crash site showed that the tank lining had been driven inward (instead of outward, as was the case with the piece found on the runway), strongly suggesting that another tire fragment or other debris had struck the tank and produced a second, much smaller hole.

After examining evidence from both the runway and the crash site, investigators had determined that there were at least two failure modes, one indirect and the other direct, leading to the burst tank. As Bouillard explains, "After these two tests, we were able to determine that the shock of a tire [fragment hitting] at an elevated speed exerted two types of pressure or forces on the tank." The indirect force that generated the shock wave likely resulted from the largest piece of tire debris that had penetrated the wing and slammed into the tank before it fell to the ground near the fuel stain on the runway. The large hole created by this outward failure was likely the source of the massive fuel leak that spilled raw kerosene, which quickly ignited and produced the flames seen trailing from the underside and rear of the plane. The second force was direct: One smaller piece of debris had struck the underside of the tank and created a second, smaller hole, as evidenced by the burned section of tank found at the crash site.

The tests conducted at the center in Toulouse had not been 100 percent conclusive, but they were convincing, says MacIntosh. "It's not unusual for investigators to have an explanation that can't be run all the way to ground." By the end of the testing phase, "it was pretty well accepted by the majority of the team that this hydrodynamic action was a good explanation of the large piece of fuel tank that appeared to have been blown out of the bottom of the number 5 tank. I personally believe it was right on the money. There are some that will always have contrary theories. Those will always exist because we simply can't exactly recreate every scenario."

Determining how Concorde's number 5 fuel tank had ruptured was the penultimate step toward the ultimate goal of the investigation: arriving at an evidence-based reconstruction of the entire chain of events leading to the catastrophic blaze that had overwhelmed the plane's operating systems and set the port wing and parts of the airframe on fire. Investigators effectively took a deep breath and prepared to tackle the problem of exactly how and where the deadly fire had begun. "In the first months of the investigation, we had established the relationship between the metal strip and the tire, and then between the tire and the bursting of the tank," says Bouillard. "Next we needed to understand what had ignited all of the fuel coming out of the hole in tank number 5."

"Little by little, we had been progressing toward actually coming up with an explanation of the phenomenon of the fire," says Torres. Given the complexity of all the variables involved, competing theories were bound to emerge. A free exchange of ideas and insights from everyone involved would be useful to establish common ground and discuss scenarios as to what started the fatal fire. The international BEA team would need all of the expertise at its disposal. "So we decided to gather together specialists from many different organizations to lead a think tank on the conditions surrounding the fire," says Torres.

There was no shortage of facts at their disposal. The seven teams of investigators had compiled thick files full of photographs, research, and reports. A consensus already existed as to findings regarding the volume and flow rate of the leak streaming from tank number 5, Bouillard says. "It was important to know the rate of flow of the fuel leak for the simple reason that a slow leak—such as a leak of 4 kilograms [9 lb] per second, as in the Washington, D.C., incident—would not lead to an ignition of fuel." By contrast, the sheer volume of fuel gushing from the ruptured number 5 wing tank was massive. Estimates put it at 60 kilograms (132 lb) per second—a rate 15 times that of the leak in the 1979 incident at Dulles Airport.

Concorde 203's fate had been sealed not simply when the fire was sparked—whatever the source—but when the fire became stabilized. The more the investigators examined what they knew about jet fuel and fires, the more they were persuaded that what had happened to Concorde was brought on by a concatenation of unusual events and peculiar conditions that all built toward creating the perfect flame.

There is a common misconception that liquid fuels burn. In fact, they burn mostly as gases rather than as liquids. Before the efficient combustion of a liquid fuel can take place, the fuel must be converted into a gaseous state. That's because the chemical oxidation reaction of combustion results from the combination of substances at the molecular level. At that level, molecules of fuel must come into direct contact with molecules of oxygen before they can combine to form the new molecules of carbon dioxide and water, which, when exposed to a flame, release the heat energy that we experience as fire.

For example, a bucket half filled with liquid jet fuel will burn if lit, but not very quickly or intensely. It will only burn at the exposed surface, where the fuel molecules come in direct contact with the air molecules that contain the oxygen required for the chemical reaction. The vast majority of the fuel molecules in the bucket remain below the surface, where they cannot come into contact with air molecules. So they have to wait their turn to burn. The result is a slow, rather feeble fire confined to the surface that gradually releases energy as the fuel burns down, exposing the remaining kerosene molecules to the surface, where they can get the oxygen needed to ignite and burn.

For jet fuel to burn at maximum efficiency, a Goldilocks-like standard must be met: Everything has to be just right. When vaporized fuel is mixed with air in the engine's combustor, the balance of fuel to air in the mixture needs to be within certain limits to burn efficiently. A mixture with too much fuel and too little air is too rich and will not burn well. A mixture with too little fuel and too much air is too lean and will not burn well, either. The optimum mixture between the rich and lean that allows the fuel to burn most efficiently is called the stoichiometric mix.

Jet fuel, composed primarily of kerosene, is thus not flammable in its liquid state. It has to be vaporized and then mixed with air—both of which occur in turbojet engines under controlled conditions. Even a massive fuel leak would not automatically result in a robust fire unless additional special conditions were also present, which is what happened in the 2000 Concorde crash.

Investigators were certain that the high rate of flow helped vaporize the fuel. The dynamics of a stream of fuel pouring out through a hole during takeoff had effectively reproduced outside the plane what fuel injectors achieve inside the combustors of a turbojet engine: accelerating liquid fuel at high velocity through a restricted orifice in order to atomize or vaporize it. The amount of the fuel in the tank created immense pressure, increasing the rate of flow. The hole in the wing tank achieved the same effect by acting as a spray nozzle, instantly atomizing the liquid fuel. Finally, the turbulence of the airflow beneath the plane helped to mix the atomized fuel with oxygen in the atmosphere. "We then had a stoichiometric mix, meaning an extremely flammable mix," says Bouillard.

The moment a spark or flame was introduced, the flash of instant combustion was inevitable. "These were the ideal conditions to ignite this mix," says Bouillard. "In cases of less substantial leaks, such as what we saw in Washington in 1979, that involved direct perforations, the leaks were 10 to 20 times less substantial, and the conditions of the mix were so weak that the risk of inflammation was almost nil." Consequently, "after their investigation, the NTSB had concluded that the perforations made for a leak that didn't allow for ignition of the fuel. So, the investigation had concluded that this level of leakage could not lead to a fire, and the authorities had decided not to reinforce the lining of the Concorde's tanks." This decision made sense at the time. None of the American or French investigators had foreseen a radically altered scenario in which a massive fuel leak would ignite a blaze capable of destroying a Concorde in less than two minutes.

The 2000 tire failure at Charles de Gaulle Airport had played out very differently, and with deadly consequences. The creation of the ideal mix of vaporized fuel and air was only the beginning of the conditions that resulted in a perfect, self-sustaining flame. The simultaneous and total failure of the landing gear to retract was a major factor in the series of mechanical mishaps that turned the plane into an unstoppable blowtorch, explains Torres. "The bursting of the tire had damaged the sensors on the doors to the landing gear well [inside the fuselage]. That prevented the gear from being maneuverable [or], very simply, from retracting." As long as Concorde's landing gear remained down, a significant drag resulted that produced "a recirculating zone of air underneath the plane," says Torres, creating conditions more akin to an eddy than a blast of high-speed air that can extinguish a flame. In the event of a wing fire, pilots are aware that slowing their aircraft will allow the fire to strengthen and spread; maintaining speed is essential and can even generate a strong enough airflow to put out a fire.

The effect of the drag had been crippling. It had prevented the plane from picking up speed at a time when the engines were already surging and losing power. It had also helped create the localized environmental conditions around the landing gear that allowed even a small fire to stabilize and spread to other areas of the plane. If Captain Marty had been able to raise the gear, the fast-moving, front-to-back flow of air beneath the plane—what engineers call laminar flow—likely would have made it difficult or impossible for an incipient flame to become a monstrous fire. "Their inability to raise their landing gear allowed the fire to stabilize," says Torres. "That's because the drag created by the landing gear persisted," as did the zones of fire-friendly, circulating air created by the presence of the large and bulky landing gear. "If they had been able to raise the landing gear, maybe the flame would not have stabilized."

But the reality was that the landing gear couldn't be raised, and the flame was sustained by both the localized air movement underneath the plane and the robust, continuous flow of fuel mixing with air in the atmosphere. "At that point," says Torres, "there was no longer any way to stop it." In his book *The Concorde Story*, author and Concorde pilot Christopher Orlebar essentially agrees with Torres, adding further clarification: "If the undercarriage could have been retracted, the fire might have put itself out. But survival would have depended upon how much damage had been done to the fuselage—which may already have been breached—and whether the number one engine could have been kept going."

All this explained how the chaos and damage unleashed by the burst tire had created the optimal conditions required to vaporize the liberated fuel and nurture an incipient flame into a full-on blaze streaming behind Concorde like a fiery tail. Still missing was the last piece of the puzzle. "What is certain is that to have a fire, there needs to be a stable flame. However, we were still missing the spark," says Torres. What had acted as the match that lit the fire that destroyed Concorde?

Three competing theories emerged, two advanced by the French BEA and the third championed by British investigators from the Air Accidents Investigation Branch (AAIB), who were also participating in the ongoing accident inquiry. Two of the theories focused on the engines and their afterburners or reheats. The third concerned a short circuit and resulting electric arc inside the landing gear bay, also called the landing gear well—a large compartment inside the fuselage where the retractable landing-gear legs and bogies are stowed after takeoff. Orlebar summed it up: "There were three likely sources of ignition—the reheat, an engine surge, or an electric spark in the landing gear bay."

The AAIB theory, which could be termed the electrical-arc theory, hypothesized that a spark from an electrical arc resulting from a short circuit started the fire. The British investigators were convinced that the fuel vapor had been ignited by a spark from a short circuit in the portside landing gear bay, directly above the burst left-main landing gear tire. In this scenario, a piece of tire or other debris had flown upward and through the open doors of the bay, damaging the wire-mesh insulation on live electrical cables that ran from the landing gear bay down to fans near the wheel brakes. That caused a short circuit that generated an electrical arc, which sparked the fire. The recirculating air in the zone beneath the landing gear bay helped create the right environment to stabilize the flame, and fire quickly spread along the undercarriage from the landing gear bay area to the rear of the plane.

The BEA-backed hypothesis, which could be termed the direct-flame theory, focused on Concorde's engines. Investigators wondered if flames had been spat out from the front of the engines during a surge. However, because FDR data showed that the engines did not surge until after the fire had ignited, that was quickly ruled out. But the direct-flame theory also applied to another, even more suspect source: the engine's afterburners or reheats. The afterburners—long, tubelike extensions of the engines—are used during takeoff to produce an extra jolt of thrust that results in a blowtorch-like flame shooting from a nozzle at the back of the engine. Afterburners have been a feature of military jets since World War II. They work by mingling oxygen with jet fuel that is squirted into the high-speed exhaust stream from the engine's turbine. The afterburners provide an additional dose of power for takeoff and during supersonic flight by adding a combustible fuel-air mixture to the exhaust.

Aviation experts being interviewed by the media, including former NTSB investigator John Nance, were asked in the days following the crash to speculate on the possible source of the flame. During an ABC News interview that aired July 30, when fresh but unverified details about the crash were emerging, Nance considered a scenario very similar to the one being explored by the BEA. "If those two tires did go, as the French seem to be suggesting . . . and with fuel spilling, bearing in mind the engines were in theory on afterburner, that would cause that huge flame."

The forensic approach to analyzing the origin and progress of the fire depended a good deal on where affected parts of the plane were located within the large triangular wing and in relation to the portside landing gear bay directly above the tires. Wherever the fuel vapor initially ignited, the flame still had to migrate from the point of ignition to the hole in the fuel tank, so that the first small, unstable flame could stabilize into a large continuous flame supported by the fuel vapor from the leak. If the flame began anywhere to the rear of the hole in the tank and spread along the undercarriage, which looked to be the case, the flame would be directly affected by the velocity and direction of air movement as the plane sped along at more than 150 knots.

The spark theory made sense in that regard, because the damaged section of the brake fan cables was located inside the landing gear bay compartment, directly above the tires and just to the rear of the hole in tank number 5. In this scenario, a fire that started in the landing gear bay or well would have had a very short forward distance to travel and would have been aided by the circular airflow in the zone around the landing gear. By contrast, an incipient blaze started by the afterburners at the very rear of the wing would have had to migrate *forward* below the underside of the wing in a zone where laminar airflow prevailed and would have approximated the speed of the aircraft itself. Under these conditions, a fire that started at the back of the plane would have been snuffed out before it could travel to the source—the hole in fuel tank number 5.

Both hypotheses, once on the table, were fair game for all the investigators. Torres, for one, was not convinced the direct-flame theory was the sounder of the two. "The problem that comes up in that line of thinking is that we have a plane that's speeding down the runway," says Torres, creating conditions that were "not conducive" to a stable flame migrating forward from the afterburners. He thought the match that lit the fire was not a flame but a spark. "The most probable hypothesis, in my view, was the one concerning a failure in the brake fans"—the damage done by flying tire debris to live cables connected to the fans used to cool the wheel brakes. "That would have caused an electrical rupture and produced an electrical arc that liberated enough energy to light the fire."

The next step was putting the theories to the test, this time at British Aerospace's Warton factory near Blackpool. "Using a laboratory belonging to our friends, our British colleagues, we recreated the possibility of a flame caused by a spark," says Bouillard. The technicians who set up the test didn't shoot a piece of tire tread at exposed live wires inside an electrical cable. "But they did, in the gear well, create a spark of an intensity that was quasi-similar to that which could have been produced by damage to a circuit by a piece of tire," says Bouillard. When a rich mixture of fuel and air was injected into the gear well compartment, which resembles the interior of a large metal box, the result was instant combustion. "We considered that the test was valid," says Bouillard. "So we decided to retain this as one of the hypotheses concerning the ignition of the fuel."

Another test was conducted to determine if a flame from the afterburners could have been the culprit. A full-size mock-up of the underside of the left wing was constructed, which included the extended landing gear, both of which were mounted on a concrete platform. A wind replicating the airflow beneath the speeding plane was generated by powerful tunnel fans used in aeronautical research. The flow rate was set at 206 knots (106 m/sec) to simulate the speed of the air outside the plane at the time of the fuel tank rupture. Jet fuel under pressure was then shot through an opening the same size as the hole in wing tank number 5. The rushing air from the fans carried the vaporized fuel beneath the mock-up of the plane's undercarriage and landing gear bay to the rear of the wing, where it was lit by gas burners simulating the afterburners. "The fuel ignited, but the flames were unable to advance up the fast-moving air stream," wrote Orlebar. "This ruled out reheat as the likely source of the ignition." Still, the BEA investigators maintained that the fire might have advanced through engine ducts inside the wing, where "slower moving air" would have allowed the forward progress of the fire.

The test results clearly favored the electrical arc theory. In the end, however, diplomacy had a role to play as well. "We decided, during the writing of the report, to consider both of these possibilities as being the likely explanations for ignition of the kerosene leak," says Bouillard. "These were two valid hypotheses, and we hadn't wanted to privilege one over the other. So we kept these two hypotheses as being the most probable causes of the fire, knowing that it was perhaps a combination of both."

Over the years, chroniclers of the crash have tended to favor the electrical-arc theory while acknowledging that absolute certainty was not possible. Jonathan Glancey, writing about the crash, was careful to summarize what *might* have happened in the wake of the burst tire to spark the fire and set the plane ablaze. "Kerosene pouring out under pressure burst into flames. The live cable to the brake cooling fans might have arced when cut by debris."

An institutional chronicler, the Federal Aviation Administration, took a more assertive stance in its dry but authoritative review of the crash and investigation published on its respected website "Lessons Learned from Civil Aviation Accidents." Referring to the tests conducted in the wake of the crash, the FAA anointed one theory as valid and dismissed the other. "Based upon the results of these tests, it was concluded by investigators that the most likely source of ignition of the leaking fuel was due to electrical arcing from damaged wires in the wheel well area, not hot engine parts as was originally suspected."

What has not been disputed was the atypical nature of the cascade of failures that led up to the only fatal crash involving a Concorde. Some of what occurred could have and should have been prevented, given the known vulnerability of the tires to blowouts and the wing tanks to penetration by tire and metal debris. Still, chance had played its role as well. "Of course, it was a completely exceptional chain of events that led to this accident," says Torres. "That said, when they asked us if it's impossible for it to happen again, the investigators all answered, 'No, no!' It's impossible to guarantee. . . . In this case, they really had bad luck. But it's impossible to say that the same phenomenon wouldn't happen again." As with so much else associated with Concorde, the crash was unprecedented. The tire failure itself "was not unusual," according to the FAA's online analysis of the accident. "However, what was unusual was the extent of the damage to the number 5 fuel tank that this tire failure resulted in, and that this failure was of a type never experienced before in the history of civil aviation."

The BEA-led consortium of investigators from four different countries had achieved their goal of reconstructing a plausible, evidence-based narrative of the frightening series of events that had unfolded with explosive suddenness, overwhelming the flight crew and the working systems of the plane itself, which had quickly burned and crashed. One of the final and most unsettling forensic tasks yet to be completed would be to review in detail what the three men in the cockpit had been confronted with—a multifaceted, high-speed emergency measured not in minutes, but in seconds.

9

Delays and Headaches

In his moving accounts of his aeronautical adventures, ace pilot and author Antoine de Saint-Exupéry displayed an uncanny ability to be at once philosophically detached and viscerally engaged with the experience of flying, its joys and perils, and what it taught him about life and death. "And there I stayed a bit, ruminating, and telling myself that a man was able to adapt himself to anything," he wrote in one of his great works, the 1939 memoir *Wind, Sand, and Stars*. "The notion that he is to die in thirty years has probably never spoiled any man's fun. Thirty years, or thirty days . . . it's all a matter of perspective." Saint-Exupéry saw clearly and unflinchingly what was close at hand, but he also looked to the stars for solace. In the process, he became one of the most eloquent chroniclers of what it was like to fly into an impenetrable fog, to crash-land in the Sahara Desert, or to be stranded in a disabled plane atop a remote, frozen plateau without hope of rescue.

The modern world in which the Concorde flight crew lived and worked would have been at once familiar and strange to Saint-Exupéry, who was born in the year 1900 and lived only until 1944, when his Lockheed P-38 Lightning vanished during a reconnaissance mission off occupied France. Although he died in the service of the Free French military, Saint-Exupéry—like Christian Marty and his fellow pilots—spent most of his working life in the field of civil aviation, making long, perilous night flights over the towering Andes and the vast Sahara as a pilot for the French air mail service. He got his start in 1926 flying for the Compagnie Aérienne Française, a small airline based at Le Bourget Airport. Paris was his off-and-on-again home, and he even worked briefly in the public-relations department of Air France.

The balky, primitive planes he flew—propeller-powered Sopwith biplanes and triplanes with few instruments—bore no resemblance to the supersonic Concorde. But he surely would have been enraptured with the prospect of soaring high enough to explore the stratosphere and marvel at the curve of the Earth. His awe of the first supersonic passenger plane doubtless would have been tempered by his knowledge that it was no more than a machine built for travel. "The airplane," he wrote, "is a means, not an end." Its higher, overlooked purpose—to reveal "the true face of the Earth"—was open to discovery by a privileged few: pilots who were, like Marty, both disciplined and thoughtful.

Saint-Exupéry certainly would have appreciated the depth of comradeship that existed among three seasoned professionals working together in the confines of the claustrophobic Concorde cockpit. And he knew what it was to watch a friend climb confidently into the pilot's seat before flying off into perfect blue skies, never to return. "For nothing, in truth, can replace that [lost] companion," he wrote. "Old friends cannot be created out of hand. Nothing can match the treasure of common memories, of trials endured, of quarrels, reconciliations, and generous emotions. It is idle, having planted an acorn in the morning, to expect that afternoon to sit in the shade of the oak."

Saint-Exupéry understood as well how quickly an ordinary flight, performed as a daily service to an employer, could become a test of both a plane's limits and a pilot's mettle. Yet even he might have been astonished by the sophistication of Concorde—a marvel of structural engineering with an Achilles heel in the form of tires that were subjected to extreme forces of weight and speed unique to the world's only supersonic passenger plane, in which so many things could go so terribly wrong in a blink of an eye. Like other veteran pilots, he was tolerant of the reality that a pilot's longevity depended a good deal upon the vagaries of fate, which might grant the largess of another thirty years of life or the parsimonious allowance of another thirty days.

•

As fall turned to winter, the crash investigators had constructed a working narrative of the crash, but their work was not yet complete. Investigators continued to sift and analyze the trove of data extracted from the enhanced flight data recorder—the plane's black box. The information gleaned from the flight data recorder (FDR) would allow them to study everything that had happened to the plane and its systems during critical periods best measured in split seconds. The next step would be to synchronize their findings with the actions taken by Captain Marty, First Officer Marcot, and Flight Engineer Jardinaud during the ongoing emergency. Any assessments the investigators made were tempered by the knowledge that their analysis, which reflected the benefits of time and hindsight, would stand in contrast to the speed with which the flight crew had had to respond to the escalating crisis preceding the crash. As in any crash assessment, the investigators had to ask basic questions: How much of what the crew had done could even be understood as decisions? Might their actions be more aptly described as intuitive responses that arose out of years of experience with flying the Concorde and other planes?

In fact, what the flight crew encountered that day was so far beyond the bounds of the expected that it had never been addressed in airworthiness regulations or training procedures, observes retired Concorde pilot Jean-Louis Chatelain. "The crew had to act in a situation that was extremely difficult to assess and was not even covered by training. So they had to draw upon their own basic and deep aviation knowledge to try to get out of this situation."

A portion of the 186-page BEA report that would eventually be written, and which would serve as the official, in-depth account of the crash, would be dedicated to forensically assessing the performance of the three-man cockpit crew. Almost all of what they had done and said was available to investigators in the form of information preserved by the FDR and the cockpit voice recorder (CVR), both of which had been salvaged from the wreckage late on the night following the crash. Both recorders would shed a great deal of light on what had happened before, during, and after the onset of the emergency that erupted on the runway after the first 38 seconds of what had begun as a normal takeoff. What would prove far more difficult to ascertain was *why* the crewmembers had reacted as they had at various critical points and which choices were actually available to them given the multiple overwhelming problems they confronted.

One of the many aspects of the crash that made it unique in terms of human interactions was the fact that the Concorde was being flown by a three-man crew. That was an unusually large one for the year 2000, when two-person crews were standard. Chatelain explains: "Until the end of the 1970s, that [three people] was a classical crew complement, just like it was on the first generation of 747s and 727s." A captain, a first officer, and a flight engineer were all required onboard such earlier planes, since precomputerized technology was not advanced enough to manage aircraft systems, but by the 1980s the flight engineer's functions gradually were being replaced by computers. "We moved to a two-pilot cockpit crew," says Chatelain, "when we were able to take advantage of this new technology—computers that were able to automatically reconfigure systems. If you lose one system, then the computer is able to automatically reconfigure to another system. But at the time Concorde was designed, it was up to the flight engineer to make [any necessary] system reconfiguration."

The three men who made up the Concorde flight crew that day were all top-of-the-line veteran pilots, with 94 years of combined experience as licensed pilots, including 14 years as certified Concorde pilots. Both Marty and Marcot earlier had flown the Airbus 300, while Marty and Jardinaud had also flown both the Boeing 727 and 737. Jardinaud was also certified to fly the Boeing 747. In addition, Marty was a veteran pilot instructor, while Marcot was a certified Concorde flight simulator instructor. "No one would be given the opportunity to fly Concorde as a junior pilot," Chatelain points out; the Air France seniority system required that all three had to be quite experienced to

Concorde's cramped three-crewmember cockpit. The pilot sat in the left front seat, almost elbow to elbow with the first officer in the right front seat. The flight engineer sat in the swivel seat close behind them, monitoring the complex systems management panel. (AUTO & TECHNIK MUSEUM, SINSHEIM, GERMANY; PHOTOGRAPH © CHRISTIAN KATH)

training for Concorde certification, a lengthy process. First officers and flight engineers became eligible for a Concorde rating only after putting in 15 years as first officer on other planes. "Practically speaking, for the captains, flying Concordes was something they achieved at the end of their career," says Chatelain.

Concorde passengers sometimes were invited to get a glimpse of the interior of the cockpit, which was a world unto itself. The flight deck, inside the needle nose of the aircraft, was quite narrow and low, offering little headroom for the three members of the crew, who thus worked in extremely close quarters. Instead of the liquid crystal display (LCD) screens seen in modern, so-called glass cockpits, the Concorde instrument panel fairly bristled with rows of analog dials and gauges, including altimeters, airspeed indicators, artificial horizons, and engine temperature indicators.

The captain and first officer sat side by side, with a center console between them; the aircraft commander occupied the left seat, and the pilot acting as first officer sat on his or her right. The semicircular windshield, made of special heat-resistant glass, consisted of a series of front-facing windows connected to two additional panels on each side. Each pilot had identical black control columns, with distinctive upside-down V-shaped handles resembling rams' horns. Both pilot seats were electrically operated and mounted on tracks that allowed them to move backward and forward.

The swivel seat in which the flight engineer sat was directly behind the pilots, positioned between the two but somewhat lower. While the pilots faced forward, the flight engineer was typically turned to the right to better survey an imposing floor-to-ceiling systems management panel that provided input regarding the performance of the plane's four engines and the dizzyingly complex fuel management system that governed the plane's center of gravity. It was the flight engineer's job to pump fuel between tanks and adjust the center of gravity to keep the aircraft in trim, flying a clean, straight line.

The flight engineer was also responsible for keeping an eye on cabin pressure and temperature and the proper functioning of the electrical system, hydraulics, wheels, and brakes. He also was in charge of monitoring an array of alarms, including fire alarms. Without a doubt, "there was much to do for the flight engineers onboard these first-generation aircraft," says Chatelain. "This also meant a different kind of internal crew relationship onboard as regards task sharing." Both pilots were dependent upon the flight engineer's vital spoken reports or callouts regarding the plane's working systems. Many cockpit conversations, inevitably, were not simple one-on-ones but three-way exchanges. Concorde crews were accustomed to this triangulated communication system, which worked well day to day. But in this emergency, in which every second mattered, the lines of communication overlapped and blurred. And the flight engineer in particular would struggle to sum up a complex problem in as few words as possible.

•

The actual day of the flight had begun with unsettled, rainy weather and missed connections for hundreds of passengers arriving at Charles de Gaulle Airport. After the morning mist cleared around 10 a.m., the Concorde flight crew's workday began, and they soon confronted a few problems unrelated to the weather. First the Air France dispatcher informed them that the plane had incurred a performance penalty: its weight would have to be reduced by 2.5 percent due to a problem with a defective pneumatic motor on the thrust reverser, which effectively acts as a second set of brakes. The thrust reverser on the Concorde ensured a smooth landing by deflecting exhaust to help slow the plane after it touched down on the runway.

One of the airline dispatcher's most important jobs was to supervise the preparation of a flight dossier for the pilots that included weather reports, a flight plan, and any weather-related restrictions in effect that day. A key part of the dossier was a sheet whose purpose was to present what was called a "forecast," or estimate of the plane's final weight based on a tally of individual weights assigned to fuel, cargo, luggage, and passengers on board. After being provided with the MTOW, or maximum allowable takeoff weight of the plane for the upcoming flight, the dispatcher would then compare this limit to the actual total weight of the plane preparing to depart. The dispatcher and everyone involved in coming up with estimates of the plane's weight understood that, in practice, it is never possible to know the exact true weight of any Concorde about to take off, because of the use of average weights. (Passengers and luggage, for instance, were not weighed on scales, but assigned average weights.) A flight departure agent used a computer program to help the dispatcher arrive at the forecast, which took into account the characteristics of each aircraft. However, the basic math—adding up the weights of fuel, cargo, and passengers—was done by hand. The dispatcher knew that every Concorde needed to stay within established but also flexible preset limits for maximum weights for both takeoffs and landings, which varied from airport to airport and day to day, depending on local weather conditions and the lengths of different runways at different airports. Estimating Concorde's final takeoff weight was an inherently inexact process, but an important one. If a plane were carrying too much weight, that could negatively affect its speed and performance getting off the ground during takeoffs; similarly, too much weight might make it difficult to slow down sufficiently to ensure a safe, problem-free landing—as opposed to running off the end of the runway. Multiple factors such as surface winds, local temperatures, the length of the assigned runways for both takeoffs and landings, and the plane's mechanical fitness all had to be considered by the dispatcher and later by the captain and flight crew.

That day, the dispatcher's calculations, performed that morning before the Concorde flight crew arrived, showed that the estimated combined weight of the fuel, 100 passengers, and all of their luggage would be pushing the limits of an acceptable weight upon landing in New York, given the newly introduced x-factor—the absence of the thrust reverser that would normally help slow the plane down when it was landing.

Before the flight crew arrived, the dispatcher and duty officer—whose names were omitted from the official BEA accident report—had strategized about what could be done to remedy the weight problem, which was tied to the mechanical problem. The dispatcher proposed that they allow the plane to take off somewhat light on fuel to reduce the total weight and arrange for a refueling stop at an airport along the plane's route. The duty officer, however, first considered trying to arrange for a last-minute repair of the defective motor on the thrust reverser—thereby removing the performance penalty and weight restriction—but later suggested simply using another aircraft or loading the baggage onto another flight to reduce the Concorde's load.

Ultimately, it was Marty's call. A refueling stop would delay the flight, as would a last-minute repair. He also likely wanted to make sure everything on his aircraft was in good working order and that the landing in New York that night would be flawless. Although the flight, originally slated to depart at 3:25 p.m., would be late (in either departing Paris or arriving in New York) no matter what he decided, Marty apparently decided that fixing the thrust reverser was the best course of action. It would solve three problems: it would permit him to make a direct flight, as originally planned; it would remove the 2.5 percent weight restriction; and it would ensure a smooth landing in New York. The captain's decision was conveyed to Air France maintenance workers, who had to get busy repairing Concorde F-BTSC. Air France mechanics soon discovered that they did not have a new motor on hand, however, and they had to cannibalize another Air France Concorde to oblige Marty. It was only one of the many delays and headaches that would arise and test the patience of the flight crew that morning and on into the afternoon.

Of course, Marty had no way of knowing that another part—a small, tubular piece of aluminum that helped keep the plane's wheels in alignment—had not been reinstalled during replacement of the axle in the bogie, or wheel assembly, for the left-main landing gear carried out earlier that same month. The missing spacer was later found on the old wheel assembly. Air France mechanics had failed to follow procedures that called for it to be removed from the old bogie and reinstalled on the new set of wheels.

What Marty would have done had he known about the oversight is a matter of conjecture, but he surely would not have been pleased, and he might have requested another plane. Changing planes was not unusual. In fact, another Concorde, registered F-BVFC, was first designated to be used for the charter flight to New York, but it had been rescheduled overnight as Air France Flight 002, which left that morning. Concorde F-BTSC was the reserve aircraft chosen to take its place.

Flight 4590 was running late, and not only because of the thrust reverser repair. The passengers' luggage, transferred from a connecting flight from Germany, would also prove problematic, since obtaining security clearance for every piece of baggage was a slow process. In the end, Flight 4590 would depart an hour and 17 minutes late, and with 122 pieces of luggage in the cargo hold—19 more than the 103 bags that were listed on the load sheet given to the flight crew.

At 4:12 p.m., the plane was at the gate, with mechanics still at work on the thrust reverser, while Marty, Marcot, and Jardinaud sat in their cramped cockpit and ran through the preflight checklist. The entire recording of their conversations and asides, which lasted 32 minutes, provides an unfiltered look at how the three men jested with one another and dealt with various difficulties as they arose. "You don't start a flight at takeoff," says Chatelain. "You have the aircraft preparation and cockpit preparation, with many things to be checked and discussed, plus briefing with the cabin crew and also coordination with the loadmaster [the supervisor responsible for checking and loading baggage]. And on the day of the accident, they had a few little problems before departing."

The weather in Paris and to the north at the airport was cool and rainy, with morning mist and unsettled low-pressure systems moving southwest to northeast, from La Coruña in Spain to Leningrad in northern Russia. By 4 p.m., however, the systems had moved on; a few clouds and humid conditions lingered over Charles de Gaulle Airport, with overall good visibility. Just prior to the Concorde's departure, light breezes of around 3 knots were reported in the area of runway 26R.

The three men began by synchronizing their watches and checking the fuel gauges, along with confirming that the plane's fire-detection system was on and its batteries had been tested. At the same time, the pilots could hear one of the flight attendants welcoming the passengers—100 in all—with a cheerful "Good day on behalf of Captain Marty."

Routine duties were seen to as the crew went down the all-important review of the plane's working systems and overall readiness. Among their tasks was selecting three critical V speeds—V1, VR, and V2min—which are synchronized with events in the departure sequence. Recommended V speeds vary according to the model of airplane. Marty knew the prescribed V speeds for Concorde, but he had some latitude to adjust them. V1, which Marty set at 150 knots, is the moment of truth, the speed at which aborting the takeoff is no longer an option. VR, which the captain set at 198 knots, is the speed at which the pilot flying the plane pulls back the control column, thus lifting the nose of the plane off the ground. And V2min, set at 220 knots, is the minimum speed required for a safe takeoff. All three speeds were always higher for Concorde than for subsonic aircraft, since robust acceleration was the key to getting the plane into the air.

Speed was not simply one of the factors that went into a successful takeoff; it was *the* most important one, and it was always uppermost on the mind of any pilot at the controls of a Concorde. "Everybody on the Concorde crew knew you always need speed," says Chatelain. "There is no doubt in my mind about this." It was Captain Marty, after all, who had once told Chatelain, then training for his certification to fly the Concorde, that there was "one thing" to remember above all else: "You need speed." Similarly, all Concorde pilots were conscious of the many things that could go wrong if the plane did not reach the preset V speeds before and during rotation. Adds Chatelain, "The first officer, Marcot, he was the one who trained me in the simulator to show me what it was like going airborne at low speeds. So you cannot suspect those fellows were not aware of the criticality of flying at low speed. It is totally unthinkable."

Preparing a Concorde for takeoff was a highly detailed process in which basic arithmetic played an important role. The dispatcher manually calculated an estimation of the plane's total weight, and the flight crew then checked this number manually. An essential component of the total weight was, of course, fuel: The Concorde needed enough metric tons of fuel to make the transatlantic flight from Paris to New York City. At the same time, the plane had to stay within "maximum-performance" takeoff weight limits, which were not set in stone but varied according to other factors, including weather, the condition of the runway, and the plane's center of gravity—the last of which was in turn affected by how its weight was distributed, then rendered as a percentage.

The actual and estimated weights of everything the Concorde took on board, including people, luggage, fuel, and equipment, were tallied to determine the total load that would be lifted upon takeoff.

Naturally, passengers were not asked to step on scales prior to boarding. Passengers and crewmembers were assigned estimated weights of 88 kilograms (194 lb) for each man, 70 kilograms (154 lb) for every woman, and 35 kilograms (77 lb) for each child. Each piece of luggage was assigned an estimated average weight of 20.7 kilograms (45.6 lb).

In a purely arithmetical sense, the passengers posed no problems. Their number was set, and their total weight easily estimated. Determining the number, and therefore the estimated weight, of their baggage would be a hassle, however, due to the screening delay in processing the luggage; each piece had to be X-rayed and then matched to the passengers on board for security purposes. The pilots could not compute a final tally of the total weight until all the bags were counted.

To cope with their frustration as they waited for the thrust reverser to be repaired and the bags to be screened and counted, Marco and Jardinaud took turns grousing about the postponement to the liftoff of their powerful bird, whose wings were effectively clipped for the moment. At 4:15 p.m., Marcot made it clear to the cabin crew that he was not satisfied with the status of an aft cabin door that was not properly closed—no small matter, since explosive decompression at high altitudes can bring down a plane. "Hervé, there's still a door open aft," Marcot prompted, addressing flight attendant Hervé Garcia.

"Yeah, yeah, he's closing it now," came the reply.

"Hello, hello!" Marcot called over the radio at 4:16 p.m., trying to get the attention of the aircraft service technician on the ground outside the plane, who was responsible for inspecting equipment and supervising the mechanics on duty.

"Yes, I'm listening," came the response.

"Where are we down there?" Marcot wanted to know.

"Well, the mechanics have finished now," reported the service technician. "They're just getting off the last . . . the last toolbox now, but the loading [of the luggage] isn't quite finished yet."

Once the replacement motor had been found, the actual repair to the defective thrust reverser had taken place in less than an hour.

But the good news—that the repair that Marty had ordered was complete—failed to appease Marcot, who vented his frustration to his colleagues. "Well, it's fortunate that we're three-quarters of an hour late. Otherwise, what would it have been like . . . That's what? Ten minutes more at least [to finish loading the bags]?"

At 4:17 p.m.—five minutes into the preflight checklist—Jardinaud was continuing to review the status of the plane's complex hydraulic systems. He brought up a series of minor issues, including a leak in the plane's hydraulic system and a blockage in the plane's air intakes. "Let's first close the air intakes that are blocked . . . and then after that we're going to start up," he said, signaling that he was almost done with his review of the plane's complicated and sensitive hydraulic and mechanical systems.

Meanwhile, Marcot took the opportunity to find fault with the ground crew, who were still loading the luggage. "As soon as they've finished with their mess," he said.

Jardinaud replied with a laconic "Yeah, and then after that we're going to start up."

At 4:18 p.m., the moment arrived for which Marcot had been waiting. "OK, it's done," came the final report from the ground crew. At the same time, Jardinaud gave another progress report. "Okay, okay. I'm closing air intakes 3 and 4."

While Marcot had been hectoring support personnel, Marty listened without comment to Jardinaud's detailed review of the plane's working systems. When it was apparent the flight engineer was almost done, Marty decided it was time to check in with the cabin crew. Not to be neglected was the mood on board the plane, where 100 passengers sat buckled in their seats with no news as to how much longer they would remain stuck at the gate. Captain Marty used the interphone—the plane's intercom—at 4:18 p.m. to speak to a flight attendant, asking, "How are things in the back?"

"Everything's OK" was the upbeat response.

"Well, I'll say something just before startup," Marty said, letting the flight attendant know he was prepared to deliver an appropriately suave and formal welcome to their passengers over the public-address system after all the problems were dealt with and Concorde was finally ready to be towed from the gate. "Yes, yes," said the flight attendant.

But there were still a few minor mechanical issues to address. "Well, the air intakes are okay. We can move the flight controls," Jardinaud told Marty, referring to the plane's hydraulically activated power flight control units (PFCs) connected to the plane's "blue," or primary, hydraulic system and its "green" backup system, sometimes referred to as the mechanical system.

"We're going to be in mechanical, but it's not serious," explained Jardinaud. The flight crew, he meant, would be using the green backup system.

"We'll have to ask the dispatcher if he has a weight estimate," Marcot reminded Marty, looking for updated and revised figures. Marty was still focusing on the flight controls. "I'm in mechanical," said Marty at 4:18:35 p.m. "Maybe that's a problem because it doesn't respond."

"No, no. It's not a problem," Jardinaud replied.

"*You're* bugging me!" Marcot said to the flight engineer in a joking tone, apparently to lighten the mood in the cockpit. The flight engineer responded in kind: "*You're* bugging me if you're not ready!"

Meanwhile, Marty continued to check out the flight controls. "Well, it's making some jumps in mechanical, eh? But maybe that's normal. So next the rudder. Well, that's working, eh?"

"The flight controls are working correctly," Jardinaud replied at 4:19 p.m.

At 4:20 p.m., the dispatcher arrived with the much-anticipated revised load sheet, which included the final tally of all the baggage on the plane. "Load sheet, sir," he said to Marty. "There were some bags which were added. I have two tons two [2.2 tons] of baggage . . . for the BRS [baggage reconciliation system] software," he went on, referring to the system used for processing and checking baggage. "There's a problem we haven't defined. I called the people who were in the BRS section. Err, the baggage was correctly labeled as 4590, but it didn't go through. So we tried to do a few things to call [and] ring a few bells. But in the end we have two tons two of baggage. . . . I have no no-shows," he said, referring to the passengers. "And I have no extra bags. And all of the bags have been through X-ray, according to required random procedure."

With the load sheet in their hands and an explanation from the dispatcher about the slight increase in baggage weight and fuel being burned while the plane was taxiing to the runway, the pilots were finally able to calculate the aircraft's estimated total takeoff weight, including passengers, crew, baggage, equipment, and fuel.

At 4:21 p.m., the dispatcher announced that the ground crew had closed the hold to the baggage compartment. "Otherwise for loading, we'll close the hold and we're ready." A minute later, he bid the flight crew farewell with a breezy "Goodbye, gentlemen . . . see you soon."

Marty and Marcot echoed the salutation.

"I've taken the takeoff weight as 185 tonnes," Jardinaud told Marty and Marcot. After reviewing the load sheet and the all-important weight of the fuel, which at 94.8 metric tons was on the heavy side, Marty made his own calculations. "Well, we'll take one hundred eighty-five one hundred," or 185.1 metric tons. "That's to say we'll be at the structural limits," he said, referring to the official recommended structural limits for the plane's total estimated weight at takeoff—185,070 kilograms—that every Concorde pilot knew by heart and considered when reviewing his plane's final weight tally. At an estimated 185,070 kilograms, Flight 4590's approximate weight was close to tipping the scales as regards weight and the plane's structural limits. (BEA investigators would later determine that the plane actually took off "slightly overloaded," with a weight of approximately 185,757 kilograms, or 687 kilograms [1,515 pounds or 0.76 US tons] over the prescribed takeoff weight.) The next step in the process: determining the plane's structural balance and setting the center of gravity.

Shortly after 4:22 p.m., Marty made his final decision regarding the plane's center of gravity at takeoff: "Structural, err—fifty-four for one hundred CG." The plane's center of gravity was set by Marty at exactly 54 percent—an adjustment that would be made by the flight engineer, who was in charge of pumping fuel between the plane's forward and aft trim tanks.

Marty's decision was made quickly, but it was based on his in-depth knowledge of Concorde and its stability requirements for takeoff. In matters of its weight and balance, the Concorde was both delicate and demanding, which suited its status as a technological ballerina. Not only did the plane's weight have to be kept within prescribed limits, but it also had to be properly distributed in relation to the plane's center of gravity. As a further complication, the center of gravity needed to change from one fixed point for takeoff to other, flexible locations during supersonic flight as a means of controlling the plane's pitch—the upward or downward attitude of the nose.

One of the Concorde's trademark features was its distinctive tricycle or three-point landing-gear configuration, with the two main landing-gear bogies on each side and the single forward landing gear in the middle of the fuselage. The optimum center of gravity upon takeoff ensured that the plane's weight was centered over the two main landing gear and bogies, with some weight over the single forward bogie. Every Concorde pilot understood that the plane's *usual* center of gravity upon takeoff should be near 53.5 percent, or slightly forward of the main landing gear bogies. Locating the center of gravity just forward of the main landing gear bogies ensured that a small portion of the weight would be distributed to the forward landing gear, so that the nose of the plane would stay down during taxiing.

Once again, the Concorde had to do something other subsonic planes did not, which was to be aerodynamically sound at both subsonic and supersonic speeds. During subsonic flight, the pitch—the upward or downward attitude of the nose—was controlled by extending the elevons into the airstream, much as other subsonic aircraft do. The elevons on Concorde were control surfaces on the trailing edge of the wing that combined the aerodynamic functions performed by the ailerons, used for roll control, and the elevators, used for pitch control, on aircraft with normal wings and tail stabilizers. However, during supersonic flight, this was no longer possible because the high-speed airstream would either rip the elevons off or create too much drag. So pitch was controlled by shifting fuel back and forth between the special trim tanks. When the fuel was pumped to the aft tank, the center of gravity also moved slightly aft, moving the tail downward and the nose upward (rotating around the center of gravity) and allowing the plane to climb—an adjustment appropriate for takeoff.

Knowing that the plane's takeoff weight was very near or over what was acceptable in terms of the plane's weight and structural integrity, Marty apparently compensated by directing Jardinaud to shift the plane's center of gravity location slightly aft. A veteran pilot such as Marty would have been aware of the advantage gained by moving the center of gravity back to the 54 percent location: It increased the amount of force the elevons would exert during rotation, allowing the plane to take off with an extra ton of fuel above the usual weight. Critical press reports that were published after the crash pointed to the excess weight and atypical center of gravity chosen by Marty as evidence of sloppy preparation, but the reality was quite different: Marty had adjusted the center of gravity to make the best of a less than ideal situation.

At 4:25 p.m., Marty himself checked one last time with the ground crew to make sure all of the plane's doors were secure.

"Okay, and you confirm that all the doors are checked," said Marty, referring to baggage compartment doors that are closed from outside the aircraft.

"The doors are checked, yes," came the response from one of the ground crew.

Shortly afterward, Jardinaud made his formal report that the checklist was complete—"The pre-startup check list has been done"—followed by further confirmation from Marcot: "Yes, yes, we're ready."

Marty could at last get on the plane's public address system and make an announcement that likely relieved the passengers: "Ladies and gentlemen. All is in good order, and we are starting up our engines." Two minutes later, one of the airport's heavy-duty towing tractors arrived to hook the Concorde to its tow bar, push it back from its gate, and position the big plane facing east.

At 4:34 p.m., the ground controller bid the cockpit crew goodbye. Shortly afterward, air traffic controller Logelin took charge of the plane's movements. "Air France four five nine zero, good day. Taxi to holding point twenty-six right [26R] via Romeo," said Logelin, clearing the aircraft to travel eastward via the long, east-to-west Romeo taxiway, then jog south on a shorter taxiway that brought the plane to its intended destination: the threshold of runway 26R.

With the plane taxiing under its own power, the three men continued with the post-startup review, which included checking the emergency brakes, front-wheel steering, and brake temperatures, as well as the status of various flight controls. For his part, Marcot was concerned about the potential for overheated brakes, saying of the brake temperature, "Yeah, it goes up on this runway. We'll have to watch out."

Concorde flight crews had to be constantly vigilant and know their planes inside and out. They also had to master themselves sufficiently to make the radical transition from the fastidious planning that characterizes preflight preparations to the adrenalin-pumping high-speed takeoff. "Flight planning was quite challenging," says Chatelain, "because we would go through three different regimes: subsonic, then transonic, and finally supersonic. And each regime required specific checklists." The moment that the checklists were completed and the captain put his hands on the throttles was always charged with anticipation. "At takeoff, the acceleration was tremendous. We would reach high speed soon after takeoff and basically fly the aircraft full speed all the way."

An essential part of preflight planning was, of course, ensuring that Concorde would not run out of fuel during the long transatlantic crossing. To guard against this possibility, Concorde pilots had to be prepared at all times to divert to airports within reach where they could refuel. "If anything were to happen to Concorde that would prevent it from making the entire transatlantic crossing in the supersonic regime, and force it to return to subsonic, then you might not be able to make the crossing with enough fuel," says Chatelain. "So we had to compute some diversion scenarios." Such events forcing a return to subsonic speeds could range from a failure of one of the plane's four engines to a problem with its air conditioning systems. In such a case, Chatelain says, "we would have to descend and fly at a lower speed with additional drag. In such a scenario, we had to make a decision on continuing the flight, making an in-flight return, or diverting to a nearby airport. So we would typically take an en route diversion to airports such as Shannon in Ireland, Gander in Newfoundland, or Halifax in Nova Scotia."

In general, making sure the plane had enough fuel to make the crossing but wasn't too heavily fuel-laden was always a balancing act, one that required close communication between the pilot flying and the flight engineer. On one memorable British Airways flight from New York to London, a flight engineer informed his captain that their Concorde was running low on fuel and recommended an en route diversion to Shannon to refuel. Determined to arrive on time, the captain ignored this advice, opting instead to press on to their final destination, London's Heathrow. "Punctuality was more of an issue than gas in the tanks on this occasion," former British Airways flight attendant Sally Armstrong writes. "Concorde did land safely at Heathrow but ran out of fuel after leaving the runway. That was a close shave. . . . Perhaps lawsuits would have been issued faster than a Concorde takeoff had the truth come to light."

Concorde required that the pilots who flew it had to be prepared to embrace the contradictions that were built into its design. It was an incredibly powerful and durable plane built to endure the rigors of supersonic flight. On the other hand, it was also a thoroughbred, both highly sensitive and difficult to master. The plane was, as author Jonathan Glancey observes, "sublime and magisterial," but susceptible to unique mechanical and logistical problems that tested the skills of its highly trained flight crews.

At 4:39 p.m.—three minutes before takeoff—Marty began his standard but essential review of the list of emergency procedures he would follow if serious problems arose during the takeoff. Like other Concorde captains, he was well aware of the plane's vulnerability to burst tires; after the close call at Dulles Airport in 1979—when the pilot had been oblivious to the burst tire that put a large hole in his wing tank until a passenger pointed it out—all Concordes had been equipped with sensors connected to a warning light on the instrument panel that let the pilots know a tire failure had occurred. It was such a common problem that procedures had been established detailing the emergency measures that should be taken to handle a blowout. These procedures were covered prior to takeoff in what amounted to a verbal run-through of what should be done in the event of an actual emergency.

"Between zero and 100 knots, I stop for any aural warning [of] tire flash and failure callout," Marty said to Jardinaud, who was monitoring the performance of the engines and other working systems on his instrument panel.

"Between 100 knots and V1, I ignore the gong," Marty continued, referring to the sound that issues when there a drop in engine oil pressure. "I stop for an engine fire, a tire flash, and the failure callout," he continued.

"Yes," Jardinaud replied by way of formal confirmation, since it was his job to call out system failures displayed on the large panel to his right.

"After V1," said Marty, "we continue on the SID [standard instrument departure] we just talked about. We land back on runway 26 right." After the plane had surpassed the decision speed of 150 knots, there would be no going back, even in the event of a burst tire or engine failure. Instead, the plane would circle back, dump fuel to lighten its load, and execute an emergency landing on the same runway from which it had taken off.

Inside the passenger cabin, the six flight attendants were likely buckled into their jump seats and priming themselves for service of the planned five-course dinner to be served during the flight, which would take less than three and a half hours. Such meals were rarely less than lavish: as journalist Patrick Skene Catling recalls of his own Concorde travels, the transatlantic hop permitted "just enough time en route for a luncheon of James Bond gourmandism, providing, for example, caviar, terrapin, pheasant, Grand Marnier soufflé, fruits and cheeses and all sorts of vintage wines or, to avoid the stress of decision-making, champagne all the way. When the Machmeter on the bulkhead indicated Mach 2, twice the speed of sound, faster than a rifle bullet," he writes, "a few of the cognoscenti [would] look up from their coffee and liqueurs with sufficient interest to applaud."

Flight 4590's passengers, 97 adults and three children, all in the varying states of excitement and anxiety that precede a globetrotting holiday, would have been chatting, daydreaming, or gazing out the small, oval cabin windows, unaware of the flight crew's rehearsal of what they would do in an emergency that could include a burst tire or an engine fire.

Charter flights such as 4590 typically included many passengers who had never flown Concorde, according to Sally Armstrong, who crewed on British Airways Concordes. "The first-timers on the charters were always so enthusiastic, as were the children who were lucky enough to be included," she recalls. "It was a pleasure to see their faces as they stepped aboard, many speechless with excitement."

By the time the plane reached the speed of sound, the flight attendants would be serving wine and cocktails. The cabin crews on Concorde were bound together by a strong esprit de corps that came from serving on an elite luxury aircraft. Among the flight attendants on board Flight 4590 was Brigitte Kruse, 49. That day, she was the most senior steward in the cabin crew, composed of two men and four women, the youngest of whom was 27. Kruse's girlhood dream was to become a flight attendant. Her passion for flying defined her, and she built her life around it. A native of Germany, she was multilingual and at home in cities around the world. Single and childless, she enjoyed spending her holidays with her elderly parents, who lived in the town of Varel in northern Germany.

The other members of the crew working in the passenger cabin that day were Virginie-Huguette Le Gouadec, 36, cabin services director, and flight attendants Patrick Chevalier, 38; Anne Porcheron, 36; Florence Eyquem-Fournel, 27; and Hervé Garcia, 32. Le Gouadec, described by those who knew her as both elegant and kind, had been taking flying lessons and had recently earned her pilot's license. Chevalier was married and had a three-year-old and a five-year-old, and was well liked by his neighbors in his commune of Bouillancy, northeast of Paris. Porcheron also worked as a dance instructor, and it later came to light that she had arranged for a portion of her life insurance proceeds to benefit SOS Village, a charity for children from broken homes. Florence Eyquem-Fournel, the youngest of the cabin crew, was an accomplished student of languages and was fluent in English and German. Garcia had been a flight attendant for seven years and had been certified as a "Concorde professional" in 1999. Like his coworkers, Garcia "loved to fly the Concorde" and "took his work to heart," according to his brother, Stéphane Garcia.

All Air France Concorde flight attendants were required to be fluent in English and a second foreign language. After being given the approval of an Air France screening committee, composed of senior flight attendants and instructors, new hires embarked on a two-part training regime that focused on both safety and hospitality. Their training covered familiarization with profiles of Concorde's customers and every aspect of service from hospitality in the Concorde customer lounge to the niceties of food and beverage service on board. Most of the instruction focused on safety, however. The six-person cabin crews were acquainted with how to stow and retrieve onboard safety equipment; how to use fire extinguishers and life jackets; and evacuation procedures that included training in how to direct all 100 passengers to exit the aircraft onto Concorde's wing and await rescue in the event of an emergency landing on water.

None of the procedures that Marty and his flight crew reviewed to deal with anticipated emergencies would prove useful as they tried to cope with a disaster that had been set in motion just minutes before, when the DC-10 in front of them dropped a sharp, bent piece of metal on the left side of same long runway that lay in front of Concorde and its unsuspecting flight crew.

10

Three Seconds of Mayhem

The presence in any cockpit of Captain Christian Marty, described by those who knew him as mentally and physically sharp but also calm and easygoing, inspired confidence and pride in the people with whom he worked. By all accounts, he was widely admired for his expertise as a windsurfer, downhill skier, and all-around athlete. Equally at ease piloting a supersonic jetliner or flying solo in a hang glider over a volcano crater, "he was an exceptionally competent man, admired and liked by everyone at Air France who knew him," according to Bernard Pedamon, a pilot and friend of Marty's. Pierre-Jean Loisel, an Air France captain, describes Marty as "a humble man, and a very professional one."

While other crewmembers relaxed during stopovers, Marty was keen to hike, rock climb, or take to the hills on his mountain bike, which he kept stowed in the cargo hold. When it came to his work as a pilot, he was "extremely conscientious," says Claude Bouvier-Muller, a retired Air France pilot and a close friend. "Just like his piloting, he prepared all his projects in the most meticulous way. When he started talking to me about crossing the Atlantic on a sailboard, I thought he was nuts. Then I realized he had thought out every facet of the problem and that he was dead serious about it." During the 37-day crossing in 1987 from Dakar to French Guiana, Marty slept tied to his board and would not allow his support boat to tow him. "I didn't want to gain a single mile unless my wrists and arms felt it," he later said. His approach to testing himself showed his more introspective side; "I am not afraid of losing, because there are honorable defeats," he once said. All his friends were convinced of his ability to stay cool in a crisis. If Marty's plane were in trouble, he would remain calm and do everything in his power to save it, says Bouvier-Muller. "He was not one to give up even in the toughest situations. He had a stronger survival instinct than most people—perhaps because of his sporting activities."

The plane's first officer, Jean Marcot, was also well liked and accomplished. He had been a member of two Concorde crews that had broken two world supersonic speed records in 1992 and 1995. He loved flying the Concorde and had turned down opportunities to fly as captain on other airplanes. Amiable and outgoing, Marcot was considered the most popular of the copilots in the small family of pilots and flight attendants who made up the Air France Concorde crews. During overnight stops in New York City, he was known for organizing group outings and dinners, and on one New Year's Eve he had celebrated by reserving seats at a Broadway musical for his entire crew.

The disaster that was about to unfold would last 121 seconds, from commencement of takeoff at 4:42:30 p.m., when Marty pushed the engine throttles forward and began accelerating down the runway, to the flight's final moment at 4:44:31 p.m. But at 4:40 p.m.—with 2.5 minutes remaining before the start of takeoff—all three men in the cockpit were expecting to once again prove that the Concorde and its crew deserved the prestige attached to an extraordinary plane flown by some of the world's best commercial pilots.

By late afternoon, local temperatures of 19°C (66°F) and overall good visibility prevailed. Just prior to the Concorde's departure, light breezes averaging 3 knots were reported in the area of runway 26R—enough to cause smoke to drift, but not enough to make a weather vane move.

At 4:40 p.m., local air traffic controller Gilles Logelin authorized Flight 4590 to line up for takeoff. Two minutes and seventeen seconds later, Logelin, who could see the plane from his post high atop Charles de Gaulle's southern control tower, gave Captain Marty and Flight 4590 official permission to commence takeoff. "Air France four-five-nine-zero, runway twenty-six right. Wind zero ninety, eight knots. Cleared for takeoff."

In the verbal shorthand used by controllers, Logelin was confirming the correct runway to use and letting Marty know that he would be taking off with the wind coming directly behind him from the east. More important, this would be a tailwind for the westward-bound Concorde. Pilots prefer a headwind, since it actually shortens the distance required for takeoff: A plane taking off into a headwind benefits from air flowing over the wings from the front to the back, providing lift to help the plane go airborne. Conversely, a tail wind moves across the wing from back to front. It provides no lift and forces the plane to pick up more speed to generate the lift necessary for takeoff. A pilot in Marty's situation would usually take the time to request permission to taxi to the other end of the runway in order to take off into a friendly headwind. But Marty did not do so. Neither did he acknowledge Logelin's transmission.

In hindsight, it certainly appeared that Marty had made an imprudent decision by choosing to disregard the controller's report of unfavorable wind conditions. Since the plane was slightly overweight, Marty would have known that taking off with a tailwind of 8 knots was pushing the limits of what his tires could handle. The speed required to get the plane up in the air was determined by two factors: takeoff weight and wind direction. The greater the weight, the faster the runway speed required. A tailwind would also require a faster takeoff speed. Given the report of an 8-knot tailwind, the only way that Marty could have stayed within what amounted to prescribed tire speed limits would have been to sit on the runway and burn off enough fuel to reduce the weight—creating yet another annoying delay in a long day already filled with them.

But like so much else that happened that afternoon, this decision, too, would pivot upon computer data measured in half-seconds. Intensive research conducted by the BEA investigators showed that the wind in the area of the airport's four runways had fluctuated from 0 to 9 knots between 4 and 5 p.m. on July 25, 2000, according to data provided by the national weather service, Météo France, which stores records of airport wind speeds that are recorded every half second. The average wind speed at the threshold of runway 26R during that time was 3 knots. Furthermore, the Météo France records consulted by investigators would later clearly show that air movement was "practically zero" at the exact time that Flight 4590 took off, according to the findings of the final BEA report.

The best evidence available to investigators indicated that the wind had died down in the seconds between the verbal report from the control tower and the moment of takeoff. Investigators would later speculate that veteran pilot Marty had trusted his powers of observation and noticed the absence of wind simply by looking at an inert windsock that would have been within his view near the threshold of runway 26L. If so, he had exercised his own judgment and taken no undue risks. However, as with many other small decisions made by Marty and his flight crew that day, the motives behind them must remain a matter of speculation.

At 25 seconds after 4:42 p.m., Marty conducted his final verbal check-in with Marcot and Jardinaud prior to takeoff: "Is everybody ready?" His question was met with a clear, affirmative "Yes" from both Marcot and Jardinaud.

Six seconds later, the clicking of the thrust levers is heard on the CVR as Marty moved all four throttles full forward, commencing takeoff at 4:42:30 p.m. This was the moment that the flight crew had prepared for, when the tedious and demanding preflight review was finally complete and the Concorde itself was at its most impressive as the full power of its noisy engines and afterburners was unleashed. As the plane's four turbojet engines roared to life, Marty, Marcot, and Jardinaud were focused and silent. But one of the flight attendants seized the opportunity to pick up the interphone and cheer on their captain with a jubilant "Go, Christian!" likely heard by all three men.

Even veteran flight attendants say that taking off in Concorde never lost its thrill. "Concorde was like a caged bird," writes former flight attendant Sally Armstrong, "As soon as she saw the runway, she couldn't wait for her release to be airborne. . . . The captain would engage the throttles and as they went fully forward, there would be a slight nudge in the back as acceleration gathered pace down the runway. Again another surge of excitement and thrill that in thirty seconds one would become airborne and soar into the skies."

During the first 38 seconds, the takeoff was uneventful in every respect. Marcot announced the plane's passage through 100 knots and the decision speed of V1 without any hint of a problem. "They start setting takeoff thrust, and they had the ignition of the afterburners, and everything was normal," says Concorde captain and BEA investigator Jean-Louis Chatelain. "The acceleration was also normal."

That perception of normality was upended in an instant at 4:43:09 p.m., when the speeding plane had traveled just 1.7 kilometers (1 mi) and the number 2 left-main landing gear tire ran over a bent titanium strip that lay on its side with its razor-sharp edge pointed upward. The result was an explosive blowout at the worst possible moment: when it was too late to abort the flight. "There was a point where they reached V1, which they called out," says Chatelain. "And then something unusual happened, something that was not covered by the airworthiness regulations, that was not covered by training. And it was something that, in a pilot's career, you don't want to face, something potentially catastrophic."

The *something* to which Chatelain refers was a series of rapid-fire events that played out over the course of the next three fateful seconds, starting with the violent blowout of the number 2 tire that hurled shrapnel-like tire fragments outward and upward. One of the fragments shot straight upward and into the number 2 wheel well, where it tore the insulation off a live electric cable. Another larger, 4.5-kilogram (9.9-lb) chunk of tire tread did the most damage, striking the number 5 wing tank, setting off an internal shock wave that ruptured its hull, while another, much smaller piece tore into the number 5 tank, gouging out a splinter-shaped rectangular opening. Thousands of pounds of liquid kerosene began pouring from the large, square hole in the fully loaded number 5 tank, gushing out into the air, where the fuel instantly vaporized as the plane continued to speed onward. All this took place in a few tenths of a second after the CVR recorded the fleeting pop or bang of the bursting tire.

One second after the tank ruptured, an electrical arc from the damaged cable in the number 2 wheel well ignited the fuel vapor, which exploded in a burst of flame and black smoke seen by horrified witnesses. The fire shot under the fuselage and backward with the power of a giant blowtorch, creating a fiery tail as long as the aircraft itself. Hot gases from the blaze were instantly sucked into the twin portside engines, along with pieces of the blown tire, causing a dramatic drop in power in the number 2 engine and surging in the number 1 engine.

At the same time, with each rotation of the wheel rim, the remnants of the number 2 tire were being dragged along the runway, generating powerful frictional forces on the port side and pulling the plane off center and to the left, exactly the way that a blowout on the left front wheel of a car will pull its chassis to the left. The change in the plane's heading was worsened by the sudden aerodynamic disturbance on the left underside of the plane as fuel escaped and suddenly ignited.

Marty would have seen the burst tire alert—what he referred to as a "tire flash" during the preflight cockpit rehearsal of emergency procedure—on his instrument panel as the plane started tracking to the left. Without hesitation, he reacted by deflecting the plane's rudder to the right to counteract the leftward shift in direction. "He had lots of rudder in to try and straighten the plane out," says Bob MacIntosh, another BEA investigator. Marty's instinctive deflection of the rudder was so quick that the plane's initial rudder deflection of 8 percent began exactly one-tenth of a second *before* the onset of engine surging caused by the ingestion of hot gases from the fire.

In the same instant, all three men were being bombarded with confusing sensory information from multiple directions. "If you try to put yourself in the cockpit, reviewing what was happening in a lapsed time of three seconds—not more—they have a puzzling half-second noise linked to the tire burst, and then the impact on the wing," says Chatelain. "Then they have this blaze that is lighting up outside. They are probably noticing strange phenomena with exterior lights that are changing very rapidly" as the plane veered to the left. "And pretty soon they have two engines that are losing power. For the crew, it was totally unusual to experience this."

As the mangled tire dragged the plane off course, First Officer Marcot shouted, "Watch out!"—perhaps alerting Captain Marty to presence of another aircraft, off to their left, the taxiing Air France 747 bringing President Jacques Chirac back from a summit in Tokyo. Or Marcot simply might have been reacting to the fact that the plane was in danger of hurtling off the runway. In either case, "he's totally surprised by what's happening," says MacIntosh, who also listened to the CVR.

The same second that Marcot shouted, "Watch out!" both go lights for engines 1 and 2 on Jardinaud's instrument panel blinked off as the plane began to slew off course. Like Marcot, Jardinaud's first aural response was a purely instinctive one: he called out, "Stop!" perhaps hoping to alert Marty to the need to abort the flight before liftoff. But it was already too late.

At 4:43:13.4 p.m.—just four-tenths of a second after Marcot shouted, "Watch out!" and Jardinaud saw the engine go lights blink off—Logelin, in the air traffic control tower, alerted the flight crew to the fact that their plane was on fire. "Concorde four-five-nine-zero, you have flames, you have flames behind you." Until Logelin relayed this report over the radio, it is likely that none of the three men had grasped that their plane was aflame.

Seven seconds later, Jardinaud made his first urgent report on the emergency in progress. "Failure eng—. . . Failure engine two," he said just as the plane's fire bell began to sound, followed by a series of loud gongs warning of the same problem. MacIntosh says of this moment: "The flight engineer is the manager of the aircraft power plants, and he's seeing an engine failing and he's announcing that."

Over the next 10 seconds, Jardinaud would see all four engines go lights blink off and on and off again. Confronted with a full-on emergency—the failure of engine number 2, the accompanying fire bell and gong, and Logelin's report that the plane was on fire from behind, where the engines were located—Jardinaud had not waited for an order from Marty, calling out, "Failure eng— . . . Failure engine two," before shutting down that engine and then announcing his action: "Shut down engine two."

Jardinaud would later be criticized for his initiative. Speculation would abound about whether the plane's surging number 2 engine should have been kept up and operating to supply much-needed power and speed, and whether Jardinaud had acted out of turn instead of waiting for Marty's direct order. However, the real dynamics in the plane, supported by evidence recorded on the FDR, reveal that Jardinaud and Marty were effectively reading each other's minds and doing exactly what they had been trained to do.

Just one second after Jardinaud announced he was shutting down engine number 2, Marty gave his first order, calling out, "Engine fire procedure." Simultaneously he shifted the thrust lever for the number 2 engine to the idle position. Unaware that the fire was coming from beneath the left wing—and had begun as a result of a massive leak caused by the tire blowout—both Marty and Jardinaud made an instantaneous decision to try to control a fire that their instruments and training led them to believe had begun in the failing number 2 engine.

Retired Concorde captain John Cook describes the scene as the flight crew might have seen it, saying that all their actions indicate that they believed an engine fire had occurred. "The first thing you know of a fire is a loud bell that rings. At the same time, a red light comes on, on what's called the fire stick. You push a button to push the bell off, because it's very distracting. And then the engineer pulls the fire handle. That shuts down the engine and should, in ordinary circumstances, control the fire."

The speed with which events were unfolding made it impossible for the flight crew to communicate in any nuanced way. "A lot of things were happening on the flight deck in not just seconds but almost milliseconds," says MacIntosh. "They had a lot of pieces of this puzzle, but they weren't all coming together in a way that made sense."

The verbal stumble in Jardinaud's first callout reflected the difficulty of conveying, in as few words as possible, the information displayed on his systems management panel. "He was obviously looking for the appropriate words to pass along the information he was assessing from the engine parameters," says Chatelain. "On the CVR transcript, it looks as if he was hesitating. Of course he was hesitating! But not because he didn't find any standard callout. No, it was because he was assessing a complex situation and trying to pass the information to the pilots, who are looking outside and not looking at the engine parameters. And so the flight engineer, I think, did very well. He tried in a few words to assess the situation and relay the information."

As difficult and alarming as the situation was, it was about to deteriorate further. The simultaneous surging in both the portside engines caused by ingestion of hot gases had resulted in a sudden, dramatic drop in power even before the shutdown of engine number 2. The revolutions per minute of portside engines nos. 1 and 2 fell to almost idle, leaving the plane at 50 percent power, with all the thrust coming from the starboard engines. The effect was immediate: The plane lurched violently to the left, and the crew would have felt themselves bodily thrown in the opposite direction.

Everyone sitting in the front of the passenger cabin would have experienced similar centrifugal forces acting on them. But the pilots likely caught the worst of it because of where they were sitting, near the point of the needle nose. "If you think of what was going on—suppressing the engine thrust on the left side while they still had engine thrust on the right—this generated a terrible and sudden yaw effect at the level of the engines," says Chatelain. The effect was many times worse at the front of the 61-meter- (202-foot-) long plane. "Because the cockpit was so far forward of the engines, it moved through a larger arc, increasing the lateral acceleration and associated [centrifugal] force on the crew," says Chatelain. Anyone who watches Formula One racing has observed a similar phenomenon when a driver rounds a curve at high speed. "You can see the driver's head and body going to the side of the car" in the opposite direction from the car itself. "They experienced something a lot like that in the cockpit."

By sheer chance, the strip that burst the number 2 tire had been dropped along the runway in the worst possible place in terms of *when* the Concorde rolled over it and burst a tire. If the strip had been dropped closer to the runway threshold, and the blowout had taken place earlier in the takeoff sequence, Captain Marty could have safely aborted the takeoff. But the ill-fated encounter between the tire and the strip occurred at a point where stopping the takeoff was no longer an option. "At this point, they'd already passed the V1 speed where they would normally reject a takeoff," says MacIntosh. "Now they were committed."

Realizing he was losing control of the plane, and hoping to avoid careening off the runway onto the grass verge, Marty pulled up on the control column and lifted the plane's nose, even though he was shy of the prescribed rotation speed. "Captain Marty rotated the aircraft at about 183 knots, or about 15 knots low, to get the aircraft in what he believed to be directional control," says MacIntosh. It was a crucial decision made at a time when there were really no good choices available. "For a reason that only he could explain, the captain decided to initiate rotation slightly before the computed rotation speed," says Chatelain. "The first officer did not challenge his decision, which is also important to note. So there was a nonexpressed agreement on this. So, to me, it means that this solution imposed itself on the crew—starting the rotation early to avoid a lateral excursion, which would have been a catastrophe."

A further complication was the Air France 747 on a taxiway about 1,000 meters (3,280 ft) ahead and to the left or south of the Concorde, according to Chatelain. The jumbo jet, with Chirac on board, was taxiing toward runway 26R. The Concorde crew would have had no way of knowing the president was on the 747, of course, but they might simply have reacted to the imminent danger of a collision with the 747. No one knows if this is the reason that Captain Marty rotated, but Chatelain thinks it was likely that Marty was trying to avoid the "greater catastrophe" that would have resulted had the Concorde slammed into that other plane.

Just as compelling was Marty's primary predicament: the immediate danger to the Concorde itself. If he didn't lift off, he would lose control of the plane and crash within seconds, killing everyone on board. Marty knew that he had not yet reached VR, but he rotated simply because he had no other choice, says Chatelain. "There was a discussion [among BEA investigators] about his decision to initiate rotation—getting the aircraft airborne at a speed that was lower than the one that was computed. On the runway edge, there was one runway light that was broken [because it had been run over by one of Concorde's wheels], and we could see in the pictures [taken during that time] that the aircraft was hardly controllable on the runway."

To learn how other pilots would react in a similar situation, the BEA decided to recreate, via a flight simulator, conditions comparable to those confronting Marty. "We put some [Concorde] rated pilots in a simulator and played out the same kind of event," says Chatelain. "Most of them explained that they had a feeling of having an imminent lateral excursion off the runway." The pilots said, in effect, "Okay, we were really convinced we were not able to control the aircraft in the lateral limits of the runway."

Part of the BEA review was to pose and answer what-if questions, as in, What would have happened if Captain Marty had not rotated? "You can always think of doing something differently in a scenario that is not covered by training or airworthiness," says Chatelain. "What if the decision [to abort] had been made instead of initiating the takeoff? Again, I remind you that they were way above this critical decision speed of V1, which was 150 knots. When the chain of events started, they were at something like 170 plus knots. And by the time they were trying to assess the situation, it was something like 180 knots."

BEA technicians got busy making the necessary computations regarding the probable outcome if Marty and Marcot had decided to abort the takeoff. Their calculations took into account the speed of the plane, the blown tire, and limited braking on the left side, says Chatelain. "The answer we got was that the aircraft would have made a runway excursion at high speed with a tremendous fire already in progress. And it would have been the same result, more or less: a total catastrophe at the end of the runway."

Captain Marty most likely intended to continue the takeoff and the climb-out, anticipating an immediate return to the airport for an emergency landing. The FDR data shows that when Marty rotated, he was not simply reacting but was aware that he had not yet reached the specified rotation speed. What he did next was a further indication of the decisions he was making and what he hoped to do: compensate for the early rotation by gathering more speed as soon as he could. "There was a proven awareness of the speed issue," notes Chatelain. "At first the captain acted very slowly at the flight controls. The rotation rate was very low. He was trying to build speed."

To do that, he also needed to eliminate the drag being created by the extended landing gear. Concorde's three sets of hydraulically powered landing gears—the small forward nose gear, with two wheels, and the two main landing gear legs and bogies, fitted with four tires each—were electronically controlled by a lever on Marty's instrument panel. At 14:43:30, he announced, "Gear on retract," and immediately flipped the lever to the "up" position to raise them. "The captain knew he didn't have a great deal of ascent or rate of climb yet, and he wanted the gear up," MacIntosh speculates. "He knew he had to accelerate to something above zero climb speed—the V-ZRC—so he was really trying to correct things in the way he had been trained to do in his simulator experience. But things weren't working out that way. The aircraft was just not accelerating."

At 4:43:27 p.m., Marcot began his repeated urgent warnings about plane's frightening lack of speed. "Watch the airspeed, the airspeed, the airspeed!" He would repeat the same loud warning 23 seconds later, at 4:43:50 p.m., no doubt shouting to be heard over the loud whooping of the plane's ground proximity warning system, alerting the pilots that the plane was losing altitude and in danger of crashing.

Back in the passenger cabin, the six flight attendants were on their own, without any direction from the pilots. Flight attendants at the rear of the cabin, or possibly even some of the passengers, would have seen smoke or flames streaming off the rear edge of the left wing and smelled smoke infiltrating the cabin. Every member of the cabin crew, who knew how a normal takeoff should proceed, would have been alarmed that the plane wasn't gaining altitude. "The flight attendants who could see the fire may have begun shouting commands first," says

Cathy Cooper, a veteran flight attendant who survived a 1977 crash-landing of a Southern Airways DC-9-31 that killed 72 people in Georgia. "The others would have joined in instantly. We're trained how to handle emergencies on our own."

During the ongoing emergency, the cockpit door remained open, and the cabin services director, Virginie Le Gouadec, may have checked on the pilots and seen the three men absorbed in their struggle to fly the plane. The flight attendants would have heard the fire bell sounding and the fire warning gong going off as well. Every member of the Concorde cabin crew, Cooper suggests, would have immediately begun preparing the passengers for an emergency landing; that is what happened when she and fellow flight attendant Sandy Purl realized that both of their DC-9's engines had quit working in midair following a hailstorm. Alarmed about the state of the engines, which had begun surging and backfiring, Cooper decided to check in with the two frantic pilots, who ordered her back to her seat. Throughout the entire emergency, both Cooper and Purl, who survived the crash, did their best to remain outwardly calm and in control. "At all points in time, they would assume the pilots would somehow get control of the plane," Cooper says of the Concorde crew. "They would have known that communicating with the crew was last on the list of things to be done." Like the pilots, the cabin crew would have remained focused on their demanding job, which was to keep everyone as calm as possible while preparing to oversee the rapid evacuation of the passengers from the plane in the event of a crash landing. If the cabin crew had enough time, they would have explained how to assume the brace position and use the plane's emergency exits, says Cooper. The last thing they would have done, once all of the passengers were taken care of, was to buckle themselves into their jump seats.

At 4:43:30 p.m., Captain Marty announced he had activated the electric controls to raise the landing gear: "Gear on retract."

Marcot and Jardinaud then shifted their focus to the latest problem: The front nose gear was up, but the two main landing gear legs and bogies remained down. Apparently alarmed by a display on the systems panel that showed the gear wasn't retracting, Jardinaud called out, "The gear," two seconds after Marty's announcement.

Exactly three and a half seconds later, seeing the problem still unchanged, Jardinaud said, "The gear, Jean," this time using Marcot's first name to urge him to again try to raise the landing gear.

Ten seconds elapsed before the clearly frustrated Marcot blurted out, "I'm trying." Nothing he was doing was working, and Concorde F-BTSC continued to lumber along, failing to pick up speed.

Marcot had no way of knowing that fragments from the burst tire had damaged sensors on the two sets of doors to the main landing gear wells. The left- and right-side doors, which are automatically closed after the right and left main landing gear legs and wheels are lowered, must be opened to allow the landing gear to be raised and stored in the gear wells on each side of the plane. But the doors were either already closed or became stuck before they could open all the way. "Because these doors were not in the correct position," says BEA investigator Yann Torres, "it prevented the gear from retracting."

While Marcot struggled with the unresponsive landing gear, the smoke detection alarm in the toilet located in the hallway between the cockpit and the passenger cabin began ringing madly in tandem with the loud engine fire alarm bell and gong, which had been shut down once by the flight crew but had resumed ringing as the fire all around the portside engines intensified. Unbeknownst to the flight crew, smoke from the fire had infiltrated the air conditioning system, setting off the smoke detector alarm and likely releasing smoke into the passenger cabin as well.

At 4:43:31 p.m.—seconds after Logelin first alerted the flight crew to the presence of flames at the rear of their plane—the air traffic controller got back on the radio to let them know the fire had intensified. "Four-five-nine-zero, you have *strong* flames behind you," said Logelin. "As you wish, you have priority for a return to the field," he continued, letting Marty know that he and the other air traffic controllers were prepared to handle an emergency landing.

Confusion reigned as the flight crew struggled to interpret the conflicting signals bombarding them. At 4:43:46 p.m., Marty was still operating under the mistaken assumption that the fire seen by Logelin had started in engine number 2. He was undoubtedly puzzled and likely annoyed by the resumption of the loud, distracting fire bell and persistent gong, because flight engineer Jardinaud had announced he'd already cut power to the engine that both men perceived to be the source of the blaze. After shutting down engine number 2, Jardinaud had pushed another button near the fire handle that released the fire extinguishers, but to no avail. Uncertain about what had or had not been done, Marty wanted an updated report. "Are you shutting down engine two there?" he asked Jardinaud.

The flight engineer replied, "I've shut it down."

Meanwhile, Marcot was acutely aware that the fully extended landing gear was creating a tremendous drag on the plane. Despite his repeated efforts, nothing was working as it should. At 4:43:56 p.m., he delivered the bad news to Jardinaud and Marty: "The gear isn't retracting."

Because the pilots were unable to spare a moment to update the air traffic control tower on their desperate situation, Logelin did the only thing he could to help the stricken plane, which was to clear other aircraft out of the way in case Flight 4590 needed a dedicated runway to make an emergency landing. At 4:44:05 p.m., he radioed the fire service leader to let him know that the plane might be returning on runway 09.

Although the Concorde had first priority for returning to Charles de Gaulle, making such a return was by now impossible. The crew lacked sufficient control over their aircraft, and by this point, the blaze had progressed. As the flames spread, the plane's portside wing elevons had likely been compromised, affecting its basic aerodynamics.

Marcot heard Logelin's transmission and made a frantic report—"Negative. We're trying for Le Bourget!"—to let him know that getting back to Charles de Gaulle was now out of the question. The flight crew had another, much closer destination in mind. Directly ahead of them was their best and only hope—Le Bourget Airport. By that point, they could probably see the well-maintained airfield with its three concrete-and-asphalt runways where the Concorde could attempt an emergency landing.

At no time during the dysfunctional flight had the plane managed to fly much faster than 200 knots or higher than a radio altitude of 200 feet (61 m). The heavy landing gear legs dangled from the undercarriage of the burning Concorde, preventing it from flying any faster. The plane's left wing was on fire and losing lift. As the fire fed by the ruptured fuel tank intensified, scorched aluminum panels began flying off the top of the left wing, and pieces of the tail disintegrated and broke free, falling to the ground.

Still Marcot did not give up. His frantic callout at 4:44:14 p.m.—"Le Bourget. Le Bourget"—captured the intensity of their last hope to reach a runway on which they could attempt a landing. Just two seconds later, Captain Marty made his own, more realistic pronouncement when he said, "Too late."

After lumbering along for almost 9.5 kilometers (6 mi) as it continued to burn, Concorde could go no farther. By 4:44:15 p.m., engines number 3 and 4 were underperforming as they ingested hot gases. At 4:44:19 p.m., Captain Marty uttered his last intelligible words, "No time," 12 seconds before the plane descended onto the small hotel below.

Captain Marty and everyone on board would have felt the aircraft's nose tilting upward as the fire-ravaged port wing began to drop, sending the plane into an uncontrolled roll. The destruction of the portside half of the delta wing and fire damage to the tail and rudders rendered Concorde F-BTSC unflyable. The only direction it could go was down. Evidence at the crash site and subsequent FDR analysis showed that the plane had landed right side up—but only by chance. After rolling to the left 180 degrees until it was fully inverted, the plane turned over one more time before it hit the ground, essentially pancaking belly-down with very little forward momentum. "It fell like a dead leaf," one witness said.

At 4:44:30 p.m., Logelin, his final radio transmissions to the Concorde flight crew unanswered, watched in disbelief as a large black cloud rose into the sky to the west, somewhere among the wheat fields and motorways bordering the nearby commune of Gonesse. At 4:45 p.m., Logelin broadcast his final official report on an event he had never thought possible. "The Concorde has crashed near Le Bourget."

11

Closing the Case

Nothing the investigators could discover would remedy the loss of so many innocent lives or repair the damage done to Concorde's formerly pristine reputation. But they could—and did—determine what had caused the crash, and they would also come up with essential recommendations intended to prevent a replay of the devastating events of July 25, 2000.

Like the plane itself, the first and last crash of a Concorde would prove to be entirely unique and controversial. It was the task of the investigators to explain, as best they could, the most crucial events that had taken place in a shockingly short period of time before disaster struck. Even before an interim report was issued in December 2000, enough evidence had been collected to construct a credible narrative of what had happened, why, and how. Where evidence was inconclusive or subject to more than one interpretation, investigators had conducted tests at multiple locations in three different countries to examine their hypotheses and arrive at conclusions, all of which would be carefully documented and included in the lengthy report issued in January 2002.

Investigators had begun by painstakingly assembling all the scattered pieces of the puzzle left at the scene of the accident and on the airport runway where the most crucial events unfolded. "As it turned out, a 200-gram piece of metal on the runway triggered one of the most catastrophic crashes in aviation history," says Air France pilot and BEA investigator Jean-Louis Chatelain. Before six months had passed, Alain Bouillard and seven working groups of BEA investigators had established how Flight 4590 had crashed. Doing so was no small matter, given everything that was at stake. "We knew we had to solve the accident and also produce a report that the German and French public would accept in the wake of the loss of their citizens," says NTSB investigator Bob MacIntosh. Also paramount was insuring that all safety issues were addressed, including making sure that the Concorde's tires would finally be made more resistant to blowouts. "Of course, flight safety was at stake," says Chatelain. "But we also owed much to those who were missing—to our colleagues. So this was important to us from that point of view also. We will never forget what happened."

In general, the investigators tended to be slow to settle on a conclusion. The world wanted answers, but the French-led BEA team would not be rushed. "Despite the grounding of Concorde, we never felt a particular pressure to accelerate the investigation to be able to get it back into the sky," says Bouillard, head of the BEA-led inquiry. "We conducted it at our pace while trying to look at all the angles in a way that would give the authorities all the elements to implement the corrective measures to put Concorde back into the sky."

While they relied primarily on the wide range of carefully collected evidence at their disposal, including the trove of data extracted from the plane's black box, they also were required to make judgments informed by years of experience. Between them, the investigators had either directly been involved with or studied dozens of other major mishaps. None of them were comparable, says Bouillard. "This accident was one of a kind. It's something we could never reproduce. So it was really an extraordinary series of circumstances and also bad luck."

It was an opinion that was shared by pilot Chatelain, who was in charge of examining the role played by the flight crew. Throughout the long investigation, Chatelain was required to repeatedly listen to the CVR tape that captured the final struggle of his three Air France coworkers. Chatelain had recently been certified by Air France as a Concorde captain when the crash occurred, and he knew all three men.

His job was to objectively analyze what they had done and why and to consider as well the problems with which they were confronted. Finally, he had to make a determination as to whether their actions and decisions followed Air France procedures and training.

Without a doubt, one of the most sensitive and demanding aspects of the investigation concerned the performance of the flight crew, all three of whom had died fighting to save the lives of the 100 passengers and six coworkers on board. When it came to what had taken place in the cockpit of the Concorde that day and what could be gleaned about the performance of the three-man flight crew, a consensus began to emerge: Nothing the crew had done—or might have done—would have changed the outcome.

Using physical evidence from the runway, accident reconstruction, and data extracted from the plane's black box, the investigators had cracked the case and uncovered the circumstantial symmetry at the causal core of the disaster. In their final analysis, which would be included in the official BEA report issued in January 2002, the investigators elucidated the three primary, interconnected causes of the crash: first the burst tire, followed by the rupture of the wing tank and the ignition of thousands of tons of fuel. The official findings were listed as three bullet-point items in the "Conclusion" section of the BEA report. A fourth contributing element was "the impossibility of retracting the landing gear," which "probably contributed to the retention and stabilization of the flame throughout the flight."

Tellingly, the three failures that caused the crash were events over which the pilots had no control, starting with the metal strip that fell on the runway. "Pilots are acting in such a sophisticated system that we do not challenge the system at each point of the flight," says Chatelain. "When you're given the takeoff clearance, you assume that everything is ready for takeoff. You don't ask yourself, 'Did they make the appropriate final inspection of the runway?' No. The system is sophisticated enough. It involves so many individuals and so on that you have to trust the system."

While the FDR provided a wealth of evidence about what was happening to the plane's engines and working systems, the CVR proved more difficult to interpret. "The crew communication was [minimal], but that was understandable because there were no standard callouts matching that unusual situation," he says. Even though what they were experiencing was unprecedented in their experience, Chatelain concluded the flight crew had acted professionally. "I was in a position to listen to the cockpit voice recording many times, at least seven times, I don't remember exactly how many. And it told me—knowing all three of them—that they were focusing until the very end on trying to find a solution."

The pilots' ability to keep their wits about themselves was no small matter, since fulfilling their duties required that they exhaust every option open to them, given what they knew at the time. One of the most problematic aspects of the crash was the core dilemma that confronted the flight crew: Captain Marty, First Officer Marcot, and Flight Engineer Jardinaud all assumed they were battling an engine fire, when in fact the blaze that would bring down the plane had started in the left gear well and spread to a large hole in the ruptured number 5 wing tank. None of the tactics they employed, including shutting down the number 2 engine and using the engine fire extinguishers, had slowed the blaze precisely because it had not originated in the engines.

Investigators too had at first suspected a fire in the engines, as had Air France president Jean-Cyril Spinetta, who stood in his office watching Concorde's fiery takeoff through a picture window overlooking the southern runway. "For all of those who were eyewitnesses to this catastrophe, and I am one of them," he said later, "the cause was an engine fire on takeoff."

Arriving at the true cause of the catastrophe had included a process of elimination, putting to rest the notion that the fatal fire began in the plane's power plants. The investigators sensibly concluded that the pilots could not be blamed for failing to ascertain in less than two minutes what it took investigators months to determine. By bringing clarity to the primary causes of the crash, the investigators were able to confidently assess the performance of the pilots during the short and incredibly chaotic flight. The three members of the flight crew had done their best to manage an unmanageable situation and stave off the inevitable. Ultimately, however, the convergence of failures at the worst possible time meant that there was nothing the three men or anyone else in their situation could have done to keep the plane from burning and crashing.

All three of the critical causal events that took place on the runway *after* the plane reached V1 speed and *before* it lifted off were failures the pilots had no ability to rectify. "We were convinced that from the moment the plane rolled over the metal strip, catastrophe was inevitable," says Bouillard. "The flight crew maintained flight in extremely difficult conditions until the wing degraded and the plane was no longer manageable. They found themselves in a situation that was tragic, which they couldn't master for long. That's what we can say on the difficulties they faced and the performance of the crew."

Conversely, investigators had determined the factors that the pilots could control, including the plane's total weight at takeoff and the decision to shut down engine number 2, had not in fact contributed to the crash. Reports in the media accurately noted that the plane had taken off overweight. That was technically correct. However, as the BEA investigation would later show, the plane's total weight was within allowable limits. Also, Marty and Jardinaud had intentionally adjusted the all-important center of gravity to account for the added weight. The center of gravity was so critical that Concorde was equipped with an alarm that would go off if the plane's center of gravity were too far forward or aft. At no point during the short flight was the alarm set off.

Both Marty and Jardinaud had been determined to shut down engine number 2 as quickly as possible and keep it shut down. But both men were reacting to the loud fire alarms that went off three times during the flight while they were following prescribed procedures that called for engine shutdown in the event of a fire. As it turned out, even if they had not opted for engine shutdown, the result would have been the same. "In any event," the authors of the final BEA report noted, "even if all four engines had been operating, the serious damage caused by the intensity of the fire to the structure of the wing and to some of the flight controls would have led to the rapid loss of the aircraft."

BEA investigators decided that the three pilots could not be faulted for their mistaken judgment that the fire had begun in the plane's number 2 engine. Nor could the flight engineer be blamed for his decision to shut down the engine *before* the plane had reached an altitude of 122 meters (400 ft)—the minimum altitude for engine shutdown recommended in Air France guidelines. Jardinaud had deviated from set procedures, but he had done so based upon his best assessment of the emergency in progress. The shutdown of engine number 2 before reaching the recommended minimum altitude resulted from the captain's and flight engineer's mistaken but entirely defensible analysis of the situation.

"The investigation showed that the crew were probably never conscious of the origin of the fire nor of its extent," according to the findings in the final report. For one thing, the pilots' outlook was limited. The frontward-facing view prevented them from seeing what Logelin could see quite clearly: the yellow-red flames streaming from the left-side undercarriage and spreading to the port wing at the rear of the plane. Furthermore, their training and instruments both led them to conclude that the fire was emanating from the engines. Both the port side engines were surging and failing, which appeared to be the direct result of an engine fire.

Was it possible that the pilots had inadvertently made the situation worse by shutting down the left engine at a time when the plane needed maximum thrust to generate the speed needed for takeoff? The BEA investigators considered that possibility as well. If the engine had not been shut down and eventually recovered from the surging caused by ingestion of hot gases, it would have made no difference. Similarly, even if the pilots had somehow determined that they were not dealing with an engine fire, that knowledge would not have saved them, says Bouillard. "As to solutions, we showed that these didn't exist." Under the circumstances, he adds, the crew did well to keep the plane in the air as long as they did. "We can all salute the performance of this crew. We now know that from the point where they ran over the metal strip, the plane was lost. It was a crew that was extremely qualified and whose performance we can't criticize."

In fact, the most egregious errors uncovered by the investigation, all of which had been widely reported in the press, were maintenance issues that led back to Continental Airlines and Air France. Investigators had determined that the metal strip found on runway 26R had fallen from the number 3 engine of the DC-10 as a result of substandard repair work carried out earlier that month at the Continental maintenance facility in Houston, Texas. The other botched repair that was uncovered during the investigation was the missing spacer, which was the responsibility of Air France and its maintenance department.

As investigators discovered, the Air France maintenance crew assigned to install a new bogie had fallen down on the job. Some in the media saw this omission as a cause of the accident. "The airline's ground staff had failed to replace a 'spacer,' a vital component of the landing gear which keeps the wheels in proper alignment, when they serviced and reassembled the plane's undercarriage four days before the disaster," wrote David Rose of the *Observer*. "Although the BEA disputes it, there is compelling evidence that it was the missing spacer which may have caused the plane to skew to the left, so forcing Marty to leave the ground too early."

The BEA investigators and the head of the BEA, Paul-Louis Arslanian, who oversaw the investigative work done by Bouillard and his team, were forthright about the omission of the spacer, which the BEA investigators had in fact discovered while examining the wreckage and revealed to the press and public. "It's true that, due to a regrettable maintenance error, the spacer had not been replaced," Arslanian said. "But our investigation shows that its absence, though it slightly affected alignment of the left gear, had no influence on the way the tires were worn, or on the plane's trajectory and acceleration." Put simply, the missing spacer did not skew the left-main landing gear bogie and drag the plane to the left, nor did it retard the plane's speed.

How could the BEA investigators be so certain? They had access to the FDR data that provided a time-based record of some 400 parameters captured on magnetic tape that reflected the performance of all of the aircraft's working systems. Tellingly, the data showed that Marty did not need to deflect the right rudder to keep the plane on the runway center line before the tire burst—indicating the left-main wheels were not wobbling or creating drag, which would have resulted if they were out of alignment. "In fact, a little left rudder was applied [by Marty] due to an easterly breeze," observes author and former Concorde pilot Christopher Orlebar, who also points out that FDR data clearly showed that the temperature of the left and right brakes prior to the tire burst were "evenly warm." In other words, before the tire burst, everything in the left-side wheel assembly was working normally. The plane was keeping to the centerline of the runway, so Marty did not need to apply the brakes on the right-side landing gear to keep the plane straight—further proof that the left-side landing gear was not out of alignment from the moment the plane began acceleration. If Marty had been forced to use the right-side brakes to counteract a leftward pull, the brakes on the right-side bogie would have been hotter than those on the left side.

•

While the subject of the missing spacer would continue to generate controversy, what was not in dispute was Concorde's need for new, more damage-proof tires and puncture-resistant fuel tanks. Although it was highly unlikely that the series of events that brought down Concorde F-BTSC would ever be duplicated, that was a risk that authorities were not willing to take. In the aftermath of the crash, Air France had immediately grounded its planes, but British Airways kept its Concordes in service until the middle of August, when Britain's Civil Aviation Authority suspended the airworthiness certificate for all seven Concordes still flying.

It was a major setback and a sign of lack of confidence in Concorde, which immediately found itself in a club to which it did not wish to belong. The last plane subject to suspension for safety reasons had been the DC-10, manufactured by McDonnell Douglas. In 1979, the DC-10—a plane whose defective cargo door latching system caused a historic 1974 crash near Paris that killed 346 people—was taken out of service by the Federal Aviation Administration for five weeks after the crash of yet another DC-10-10 killed all 258 passengers, the plane's 13 crew members, and two people on the ground in Chicago.

The stakes were even higher for Concorde, and for aviation in general, simply because Concorde was so rare by comparison. In the immediate aftermath of the Gonesse crash, comparisons were made between the accident and the Hindenburg disaster of 1937, which had marked the end of the airship era. Was the three-decade-long experiment in commercial supersonic travel effectively over? France's largest newspaper, *Le Figaro*, was initially pessimistic in the wake of the crash. "Without doubt, Concorde died yesterday at the age of 31" was the verdict issued by its editorial on July 26, the day after the crash. "All that will remain is the myth of a beautiful white bird." Newspaper writers in London adopted a similarly elegiac tone: "Nothing will ever be quite the same again," wrote the *Times*. "This was the super plane, the symbol of progress, the icon of invention, a totem."

But Concorde's loyalists were legion and unwilling to concede defeat. "To the optimist, historical precedence gave more hope," recalls Orlebar. "The will had been there to remedy the fault in Comet 1 for it to return as Comet 4. A similar spirit had been shown in the case of the Space Shuttle returning to service following the *Challenger* disaster."

A consensus began to emerge that Concorde could indeed make a comeback if the necessary safety improvements were carried out. British Airways, which, unlike Air France, did not need to battle an army of compensation claims and lawsuits, was eager to get its small but profitable flagship fleet back in the air. It had signaled its commitment to that goal by taking advantage of the imposed downtime to undertake a cabin makeover and refurbishment that cost £14 million and included new blue leather seats. However, if Concorde were to take the skies again, it would need more than new seats. Improved, blowout-proof tires and reinforced wing tanks were at the top of the list of essential safety improvements being seriously studied long before the official investigation was complete.

Realizing that the survival of Concorde was at stake, advocates were hard at work behind the scenes. In November 2000, industry insiders gathered south of London at Gatwick Airport to decide what needed to be done to restore public confidence and convince authorities to reinstate the plane's certificate of airworthiness. Attendees included engineers, pilots, and executives from British Airways, Air France, Airbus UK, and Airbus France, which made parts for Concorde. The select group included six key players: Jim O'Sullivan, technical and quality director for British Airways; Captain Mike Bannister, British Airways chief Concorde pilot; John Britton and Alain Marty, chief Concorde engineers for Airbus UK and Airbus France; Hervé Page, Concorde engineering manager for Air France; and Roger Holliday, chief airworthiness engineer for Airbus UK.

Using a flip chart, O'Sullivan listed out the most pressing problems that had to be addressed. They included replacing the Goodyear tires then in use on Air France Concordes with a new, puncture-resistant tire being developed by Michelin. "The key plan of action they produced was almost startlingly straightforward," writes Jonathan Glancey. It included "Kevlar synthetic-fiber linings for the fuel tanks and new Kevlar-reinforced Michelin NZG (Near Zero Growth) tires that were nearly immune from disastrous failure when damaged."

Neither airline was required to wait for the final results of the ongoing BEA investigation to begin making necessary modifications. And both airlines had been given early guidance as to what would be required to make that possible. "We told them there are three general problem areas: reinforcing the plane's structure, mainly the fuel tanks, preventing such a fire, and ensuring the engines are not affected by fire when there's a fuel leak," said Gérard Le Houx, an official for France's Direction Générale de l'Aviation Civile (DGAC), the equivalent of the U.S. Federal Aviation Administration. "Once they give us those solutions, we can consider restoring Concorde's certificate."

Some 100 engineers began work on Concorde's revival. "What we had to do was somehow break the chain of events that occurred in the crash," said Howard Berry, an official with BAE Systems, a UK-based aerospace firm involved in the international effort to get a jump on improvements needed to satisfy air safety concerns. Work was begun fitting Concorde fuel tanks with liners made of flexible Kevlar, the same material used in bulletproof vests. Each aircraft would require some 124 of the liners for their system of large and small fuel tanks. "There was no other way to do the job other than getting the smallest and trimmest engineers to crawl into the tanks and to install the liners as if they were laying carpets," writes Glancey. But all of that work would be for naught if Concorde customers stayed away. Both airlines were highly conscious of the need to reach out to loyal customers, and they went so far as to offer inspection tours of the hangars while the refitting was underway.

The proactive work done on the vulnerable Concorde wing tanks before the BEA investigation was complete indirectly raised another issue—that of inherent flaws built into the plane's design. This was not lost on anyone familiar with Concorde's history of blowouts. "As we discover in almost every aviation accident, the causal factors are indeed a chain of events," says Bob MacIntosh. "This particular chain of events most probably started back with the design of the aircraft and the underside of the wing, and also the kind of material that was used for those fuel tanks." In the case of Concorde, not enough was done to aggressively correct chronic problems related to the tires and wing fuel tanks, despite what MacIntosh called "the hints that the gods of aviation were giving us from 1979 through 1981 when we started seeing fuel leaks and damage to the aircraft." The recurrent cause of these incidents—tires that burst and holed the big, fuel-laden wing tanks directly above the wheel assembly—had become a systemic one that was managed but never resolved.

Developing tires worthy of being fitted to any passenger plane was a major undertaking involving research and testing to ensure the end product can withstand extremes of temperatures and pressures that far exceed those required of automobile tires. "Aircraft tires can easily be taken for granted," observes aviation writer Joe Escobar. "When one considers the forces that aircraft tires endure, it seems amazing that engineers were ever able to develop these special products. Some tires are subjected to speeds as fast as a race car while at the same time supporting more weight than the largest land-moving machines." Discussion of aircraft tires by industry insiders tends to stress the same theme: the hardworking but overlooked tire on which so much depends.

Luckily, Concorde was not the only plane in need of a superior tire suitable for heavy aircraft. Michelin North America was hard at work engineering a tire tough enough to carry the weight of what would become the world's largest passenger plane, the double-deck Airbus A380, designed to challenge Boeing's dominance in the jumbo jet market. Furthermore, major manufacturers who made parts for Concorde also had a vested interest in a Concorde comeback. Leading the charge was Airbus—then known as the European Aeronautic Defense and Space Company (EADS). It wasn't long before EADS came to Michelin in hopes of finding a solution to Concorde's tire problem, which had to be dealt with in order to satisfy regulators as to Concorde's airworthiness. The answer came in the form of Michelin's new Radial NZG tires.

Those tires were just right for Concorde. The key material that made them stronger and far more puncture resistant was Kevlar. The new tires could carry more weight and were far more resistant to damage—specifically cuts inflicted by runway debris. The name Near Zero Growth was a nod to the fact that the new radial tire did not grow or expand significantly when subjected to inflation and centrifugal force. "The outer surface of the tread is not under tension," explains Orlebar. "This makes it less vulnerable to foreign object damage—it is easier to cut an elastic band when it is being stretched than when it is not."

The new NZG radial tire was radically different than the old bias-ply tires made by Goodyear and Dunlop in many respects, not the least of which was that it could not be scalped of its tread in the event of a blowout. If an NZG tire did fail, it would simply disintegrate into thousands of small, harmless pieces so tiny that no individual piece would ever exceed 1 percent of the tire's total mass. In other words, the new NZG tire offered the airplane tire equivalent of laminated safety glass that, in an accident, fragments into thousands of harmless bits instead of dangerous shards.

Eight months after Airbus came knocking at its door, the Michelin Group had developed and tested the NZG tire specifically to meet the stringent demands required for Concorde. Before trials were undertaken on a Concorde, both the old bias-ply and new radial tires were tested by Michelin and by the Centre d'Essais Aéronautiques in Toulouse to see if the NZG tires could perform normally after running over a strip of metal similar to the one that had caused the 2000 accident. In one instance, a cross-ply tire exploded after running over a strip, sending pieces of tread flying and damaging the test equipment, which had to be rebuilt before tests could resume. By contrast, the new Michelin NZG tire remained inflated, even when it was punctured by a 30-centimeter (12-in) blade.

In May 2001, the Michelin NZG tires were fitted to an Air France Concorde and taken for a spin at a flight testing center in the commune of Istres, northwest of Marseilles. By early June, Michelin was holding press conferences extolling the virtues of the new, stronger tires. "This tire does not blow up," said Pierre Desmarets, director general of Michelin's aviation tires division. On July 17, a British Airways Concorde fitted with fuel tank liners took off from Heathrow Airport for a test flight that lasted 3.5 hours. The British press was invited to witness the demonstration, which ended with a smooth landing on a rain-soaked runway at a Royal Air Force airfield in Oxfordshire. "Concorde is back where she belongs—Mach 2 and 60,000 feet," announced a smiling Bannister after he emerged from the cockpit. Executives at both airlines predicted that Concorde would be ready to fly again by the fall of 2001.

By the end of August 2001, BEA investigators were ready to issue another interim report of their findings and follow it up with vital safety recommendations. "After the report came out, we issued a list of recommendations, knowing that some modifications were already in effect, such as the reinforcement of the tanks," says Bouillard. "We also issued recommendations concerning the certification of tires and better surveillance of runways," including using radar to detect all foreign objects on the runways. Armoring of electrical wiring in the undercarriage bay was also called for to prevent damage that could result in a short circuit like the one that had produced the electrical arc thought to have sparked the fire.

Investigators knew the suspension of Concorde's airworthiness certificate had been a "major blow" to both airlines, says MacIntosh. "Our thinking was, 'Let's see what we need to do to repair the airplane and get the fleet back in the air.' That was a major effort, and it was undertaken with gusto." Regulators signaled that they were satisfied with the progress made on all fronts; on September 6, 2001, Concorde's certificate of airworthiness was officially reinstated.

News that Concorde would fly again was welcomed by everyone who worked at Air France. Chatelain recalls, "It was a kind of relief that we were able to get it back to a normal level of airworthiness. We showed the world that we were able to work on something challenging and take into account the latest safety requirements to put the aircraft back into the air."

For their part, BEA investigators were satisfied that the research they had begun had led to something that would benefit not just Concorde, but the entire aviation industry. "We now have design improvements in aircraft tires that are tenfold better than they were as a result of the research that went into the development of Near Zero Growth tires now used on major aircraft," says MacIntosh. "So we've got some takeaways from the Concorde accident that I think the traveling public should feel assured have been good things that came out of the investigation."

The stage was finally set for the resumption of the revamped Concorde's transatlantic service. On November 7, 2001—with tandem flights from both London and Paris to New York reminiscent of its twin maiden commercial flights of January 21, 1976—Concorde made its triumphal return to applause and salutes both solemn and celebratory. "This is the greatest tribute we can pay to the 113 people who lost their lives, and to whom I dedicate this flight," said Air France Chairman Jean-Cyril Spinetta, addressing banks of television cameras lined up to cover the 10:47 a.m. takeoff at Charles de Gaulle Airport.

Members of the British press were out in force at Heathrow Airport to chronicle the invitation-only London–to–New York flight of Concorde Alpha Echo, which carried 90 passengers, among them dignitaries, journalists, and aviation industry executives. The *Guardian* covered the story with panache: "Pilot Mike Bannister eased back the throttle and Alpha Echo roared down the runway. After takeoff, its world-famous nose slicing through the clouds, the unmistakable sound of the aircraft rang out."

The Air France Concorde from Paris was the first to touch down, landing at 8:30 a.m. at John F. Kennedy International Airport and beating the British Airways Concorde by 20 minutes. Mayor Rudolph W. Giuliani made an appearance to officially welcome both flights to New York City and tie them to the city's post-9/11 recovery, saying, "Welcome home, Concorde. You were missed. . . . Concorde's return is symbolic of how all New Yorkers feel about rebuilding this great city." Then he went on to urge the passengers to spend freely during their stay.

Among the celebrity passengers on board the British Airways flight was rock star Sting, who had been flying Concorde for two decades. He said he was pleased to see the supersonic jetliner resume service. "I'm still excited about going on Concorde even after all these years," he told a reporter for the *Guardian*. "Flying at twice the speed of sound gives you a buzz."

After the terrorist attacks of September 11, 2001, transatlantic travel had declined by 30 percent, bringing about what the *New York Times* called "one of the worst slumps in aviation history." (As a token concession to safety concerns, the fine silver usually provided on Concorde was replaced with plastic cutlery.) What most saw as a comeback for Concorde could help reverse that trend, observed Sting. "People are getting their confidence back slowly, and I think events like this will help people," he said. "Some people are too afraid to leave their homes and need to be encouraged to travel." British Prime Minister Tony Blair did his part to help restore confidence in the aircraft by announcing his plan to fly a chartered Concorde to Washington for talks with then-President George W. Bush.

The resumption of regular six-day-a-week flights was welcomed by those working for British Airways. After engineers conducted last-minute security and maintenance checks, a team of 270 ground staff waved off Concorde Alpha Echo in what the *Guardian* described as "an emotional scene" as the plane left its hangar. British Airways chairman Lord Marshall told the press that the airline had plans to eventually resume the normal, twice-a-day London to New York service, but that would depend on the marketplace.

The whirlwind of publicity and star turns ensured that Concorde's comeback was as dramatic as its trademark takeoff. "The return of the Concorde to service was given all the publicity of a triumph over adversity," according to the *New York Times*, which also offered a more pragmatic take on the day's events. "On Air France flights, fresh lobster and petits-fours will be served. Whether this will bring back customers is unclear. The sluggish economy and post–September 11 safety concerns might quell enthusiasm for the expensive jets."

For their part, the investigators were pleased but somewhat anxious. "We all felt that the lifting of suspension of the airworthiness certificate for the Concorde was indeed a great day," says MacIntosh. "But we all had a certain trepidation, because Concorde was 1965 technology and 1970 design. So it was just on the edge of the envelope of performance." Concorde returned to the skies as a safer plane. But *l'oiseau blanc*, the white bird, which remained a symbol of French honor, would encounter other difficulties as it flew on into the 21st century.

None of the essential and long overdue safety improvements had made it more fuel-efficient or economical. It still burned a ton of fuel for every seat occupied on a transatlantic crossing. Roundtrip tickets still cost more than $12,000. The profits it generated for British Airways and Air France were often described as slim. And manufacturers had to be convinced that it was worth their while to keep producing parts for the two small, aging fleets. Neither would the resumption of service undo the damage done by the horrific crash, which would be adjudicated in both civil and criminal courts in France in the years to come, stirring up painful emotions and initiating another fierce round of arguments about who and what was to blame for the disaster that claimed 113 lives.

Thus there was yet another unwelcome surprise in store for Concorde customers and proponents of supersonic passenger service. Less than two years after its revival, Air France and British Airways, which both went to great lengths to keep the regal and resilient Concorde in service, would make the stunning announcement that they intended to permanently retire and mothball the world's only supersonic passenger plane. It was a decision that brought no closure, but instead incited more controversy and a renewed debate about the legacy of Concorde and the future of supersonic travel.

12

Final Farewells

During its 27 years of commercial service, Concorde never lacked for fans. Perhaps the highest compliment that could be paid to Concorde was that those who loved it most were those who knew it best: its pilots and cabin crews. Veteran British Airways Concorde pilot Jock Lowe has aptly described it as "the most bewitching of machines that reduced the world to a journey of 24 hours." "It's a magic aircraft," enthuses Joelle Cornet-Templet, chief flight attendant of the Air France fleet. "The pleasure of flying it is almost a carnal one."

In general, Concorde devotees tend to look beyond the airplane's outsize development costs and nonexistent profits and emphasize instead its historic contribution to the field of aviation. Their philosophy is not far removed from the musings of Shakespeare's Lord Gloucester, who, in *Henry VI, Part 2*, compares the aspirations of ambitious noblemen to the instincts of the best hunting hawks: "My lord, 'tis but a base ignoble mind / That mounts no higher than a bird can soar." Concorde's defenders scorned the marketplace-driven concerns that would lead to its demise. "Nobody will think about whether it was a commercial success or not," said Lord Jeffrey Sterling, then chairman of the P&O shipping company, as he was about to embark on his final flight aboard a Concorde earmarked for retirement. "They will say this is another frontier which the human race has broken through."

Yet the 21st century would not be kind to the airplane, which was a product of mid-20th-century optimism, a sentiment that had fostered a vision of a technological future without limits. Photographer Wolfgang Tillmans, who captured Concorde's swanlike profile in a book of lush photographs, aptly describes what gave rise to Concorde: "A desire to overcome time and space through technology." He also muses upon what Concorde had become by the year 2003: "a super-modern anachronism." Concorde's designers had built a plane that vaulted aviation technology ahead by decades and twice set records for round-the-world flights. But it had adapted poorly to the demands of a marketplace that prioritized fuel efficiency and affordable tickets.

Still, Concorde had performed reliably and safely until the crash of July 2000 exposed previously documented weaknesses in its design, specifically the vulnerability of its fuel tanks to tire blowouts. The resulting refits and safety improvements were substantial. But they could not alleviate what amounted to a collective case of the jitters at Air France when Concorde resumed flying. "After such an event, any small daily malfunction [on Concorde] that is likely to normally occur in the aviation business would bring about much nervousness," says Concorde captain Jean-Louis Chatelain.

Among the malfunctions that cropped up after Concorde's return to service was a recurring problem with the rudder. Sections of laminated aluminum alloy skin sometimes tore away in flight. The first incident of rudder delamination had taken place in 1989. To prevent it from happening again, each plane was fitted with a new rudder, and the fix held for 13 years until November 2002, when a British Airways Concorde lost a portion of its lower rudder. Four months later, in February 2003, an Air France Concorde experienced similar in-flight damage, followed by another incident in May 2003, when a different Air France Concorde lost part of its rudder. These incidents were distressing but did not constitute in-flight emergencies: Although each Concorde has only one rudder, the rudder has two independent upper and lower sections that can function independently.

But that did not alter the fact that it was unacceptable to do nothing while Concordes continued to shed exterior parts as they sped along at twice the speed of sound. Fixing the problem would not be simple or inexpensive, since it would require Airbus to make new jigs—special tools needed to manufacture tailor-made parts for Concorde. The prospect of a costly update did not bode well for Concorde's future. And the rudder failures, which were reported in the media, had drawn fresh scrutiny concerning Concorde's continued viability. "Despite the number of people coming out in support of Concorde, they could not excuse a series of events that at best looked untidy and at worst heralded future failures," observes author and Concorde pilot Christopher Orlebar. Like a still-great athlete being treated for recurring injuries, Concorde was showing its age.

Sandwiched between the rudder malfunctions was another eyebrow-raising incident in April 2003 involving a chartered Concorde carrying British chancellor of the exchequer Gordon Brown and other officials to that year's G7 summit in the United States. En route over the Atlantic Ocean, one of the aircraft's four engines began surging. The flight arrived 30 minutes late, and the engine problem was widely reported by the press. Engine surges during supersonic flight were rare and therefore troubling. Almost as troubling was the knowledge that every mechanical incident involving a Concorde carrying high-profile passengers inevitably would make the news. "To endure another major incident, or even a series of minor ones, threatened Airbus's reputation," writes Orlebar—thus forging what amounted to "another nail in Concorde's coffin."

The rudder problem and the engine surging incident together raised questions about how much it would cost to keep Concorde in top condition for another decade. Both fleets were due for a major overhaul, according to engineers at Airbus, who indicated that the cost would be prohibitive—£40 million for British Airways alone. In addition, Airbus wanted more money to maintain Concorde. "Airbus, with its reputation on the line and a new aircraft imminent, was not prepared to maintain support of Concorde at the old price," says Orlebar. Air France was cool to the prospect of paying more for operating Concorde at a time when sales were lackluster. That left British Airways shouldering the entire cost of maintenance, which it was not prepared to do.

As the end drew near, alarmed British Airways Concorde pilots rose to the plane's defense. Concorde captain Les Brodie insisted that Concorde pilots were prepared to vouch for the viability of its airframe and the functionality of both its systems and their backups. "Having flown Concorde on test, where emergency systems are exercised, Brodie's declaration of faith is a telling one," observes Orlebar. A well-intentioned plan to maintain a token Concorde presence in the skies over Britain and to mute inevitable criticism for retiring Concorde was launched by Brodie and other Concorde crew members determined to save the plane. One proposal involved establishing a Heritage Flight program that would retain a Concorde to be flown on special occasions, such as Queen Elizabeth's Golden Jubilee flypast on June 4, 2002. But even this slimmed-down plan failed to win support from Airbus.

Airbus had decided that it was no longer worth its while to devote resources to producing parts and supplying ongoing maintenance support for Concorde and set a cutoff date of October 2003. The decision dealt a death blow to Concorde.

Less than two years after Concorde aficionados hailed its return to regular transcontinental service, the world's only supersonic passenger plane was on its way to a staid retirement as a museum exhibit. In April 2003, both Air France and British Airways, citing rising maintenance costs and low passenger numbers, jointly announced Concorde's pending retirement. Few were surprised by the decision, including Chatelain. "It was an aging aircraft, and we knew that," he says. "It had to stop one day. On the personal side I told myself, 'Okay, you had a chance to get to know the technology, but you will never again fly the aircraft. This is life.'"

Many Concorde buffs were less than sanguine about the decision made by executives at Airbus and the two airlines. "Imagine a world where Porsche came out with its 911 series shortly after Henry Ford invented the Model T, only to take its sleek roadsters off the street because they became expensive to maintain and appealed to only the most elitist of drivers," wrote Edward Wong of the *New York Times*. Leading the charge was Donald L. Pevsner, an aviation attorney and Concorde tour operator. Pevsner had flown Concorde dozens of times and organized numerous chartered excursions, including two round-the-world flights in 1992 and 1995; the 1992 flight set the westbound global circumnavigation speed record, and the 1995 flight set the eastbound record. Pevsner was a close friend of Concorde pilot Jean Marcot, who acted as copilot for both record-setting flights. He argued that Concorde would have been technically safe and sound for another decade, making its retirement premature. He blamed the decision on a combination of "cowardice" and "corporate politics" in the board rooms of Air France, whose executives remained deeply uneasy about the possibility of another accident.

Others were not as confident that Concorde would remain problem-free for another decade. "It was honestly facing an unsafe condition that was discovered through an accident," says Chatelain. "I've been involved in this flight safety business and accident investigation before, so that was the logic of it. That was not to be challenged."

Author Jonathan Glancey argues that Concorde had become anathema to airline industry executives infatuated with newer aircraft that were "crammed with every digital gizmo going." Like other airliners, however, Concorde's fate would be weighed on scales that balanced profit and loss. "The costs were rising, maybe sometimes unnecessarily," says Chatelain. "But this is the way our modern life is—we are ruled by stakeholders. The operations were complicated and the economic situation was unfavorable." The realities of air travel in the wake of the terror attacks in New York City and Washington, D.C, had made it difficult for Concorde to reclaim its niche in the depressed travel market. Passenger numbers were down as more business travelers opted for cheaper flights. "After the unfortunate events of September 11, 2001, and after a new look at the finances that were required to keep the Concorde fleet in the air, it was not too surprising that by 2003 it might be time to put the aircraft into static displays," says National

Transportation Safety Board investigator Bob MacIntosh. "It was a sad day when we saw the retirement of the aircraft and began to recognize that it was going to become a museum piece. On the other hand, recognizing that it was 1965 technology, it was time to retire the aircraft and go on to other, more modern things."

Seldom dwelt upon but always in the background was the stigma of tragedy left by the crash of July 25, 2000. "The crash was of course an element that was important in the final decision," says Chatelain of Concorde's retirement. "With such a small fleet, you cannot afford a significant [adverse] event, not to mention a crash." Air France had yet to receive the final bill for the crash: Compensation lawsuits were pending, as was the more daunting question of criminal liability, which would be decided by French officials conducting a judicial review of the accident.

On Friday, May 30, 2003, the last Air France Concorde flew out of Charles de Gaulle Airport on its way to New York City's John F. Kennedy International Airport; onboard the historic flight were 58 passengers, along with eight cabin crew and three pilots. "It's very emotional," remarked Jean-Pierre Lefebvre, an Air France staffer, before Concorde took off. "Concorde is a story of joy, of emotion, of technical prowess." On Saturday, as the Air France Concorde, flown by Captain Chatelain, winged its way back to Paris, a chartered Concorde took passengers for a spin around the Bay of Biscay. Many were in tears when they disembarked. "In France, we don't know how to hold on to what is beautiful," said one sorrowful passenger.

Air France Chairman Jean-Cyril Spinetta tried to strike a more positive note, saying, "Concorde will never really stop flying, because it will live on in people's imagination." Most of the rank-and-file Air France family members were openly mournful, however. "We feel we have all been orphaned," said Sebastian Weder, the team commander on the final commercial flight. "It was our life—every day of it—and from May 31 there will be a great void."

The British were not to be outdone when it came to ceremony and sentimentalism. The date chosen for Concorde's farewell was historically significant. The unique supersonic aircraft that took its inaugural test flight in 1969—the same year that astronauts set foot on the Moon—would make its final commercial flight on the centenary of the Wright brothers' first controlled powered flight.

Every honor would be paid to Concorde on the day it made its final curtain call. The farewell flight of the plane the British fondly referred to as "Speedbird" and "Pocket Rocket" was a widely publicized event that drew large crowds. Airport management erected a grandstand for 1,000 spectators. Some Concorde fans went to great lengths to secure a good seat, arriving hours before Concorde was scheduled to touch down. Spectators included John Cowburn, 39, from Basingstoke, a Concorde buff who brought a ladder to make sure he got a good view. "Today is a very sad day," he told a reporter for the BBC. "But we must make the most of it. Concorde is potentially the most special thing man has ever built."

The plan called for a British Airways Concorde to complete its last scheduled flight along with two chartered flights. The three planes landed within five minutes of one another at Heathrow Airport. One had flown in from John F. Kennedy International Airport in New York, while another, from Edinburgh, Scotland, was carrying winners of a competition whose prize was a seat on the supersonic jet. The third Concorde took passengers on a looping pleasure flight from London's Heathrow out over the Atlantic Ocean to the Bay of Biscay, just west of France, and back to Heathrow. The Concorde that left New York at sunrise was given a colorful salute on the runway at JFK Airport; water cannons were brought out to spray the plane with jets of water tinted red, white, and blue—the colors of the flags of France, Britain, and the United States.

Passengers on the flight from New York were special-invitation guests who included actress Joan Collins, veteran broadcaster and frequent flyer Sir David Frost, and model Christie Brinkley. "I couldn't resist one more chance to just pop over to London," Brinkley said before boarding. The mood in the cabin was emotional as the plane touched down, eliciting "tears and cheers" from passengers, according to Collins.

The New York–to–London flight was piloted by British Airways' chief Concorde pilot and captain, Michael Bannister, a minor celebrity in his own right, who told reporters that he had logged 8,000 hours and 8 million miles on Concorde since 1997. "There's a little sadness," he said. "What we've tried to do is make the retirement of the Concorde a celebration. It's something we'd like to do with the style and grace and elegance befitting this majestic aircraft."

Among those with mixed feelings about Concorde's withdrawal from service was chief BEA investigator Alain Bouillard, who cherished his memories as a Concorde passenger. The plane itself was "simply magnificent," he says. "Today it is just a marvelous souvenir stained by a tragedy." But not all news reports were filled with touching moments and heartfelt farewells. The *New York Times* sent its reporter Clyde Haberman to Queens to cover the final departure of Concorde from JFK. He found that the residents of nearby Howard Beach were relieved to see the last of the "excruciatingly noisy" aircraft. While other reporters were jotting down the words of celebrities, Haberman served up his savvy take on the contrast between the tears of Concorde buffs and the sneers of working-class New Yorkers. "There will be toasts to faded glamor and reflections on how man's eternal yearning to go ever faster, ever higher, will now go unfulfilled," writes Haberman. "But there is at least one place where none of those things will happen. It is Howard Beach. For more than 25 years, that patch of Queens along Kennedy Airport's western flank has deemed the Concorde a mortal enemy, a plaything for the well-heeled that it paid for in rattled walls and jangled nerves."

Among those Haberman interviewed was Howard Beach resident Betty Braton, who said she never got used to the plane's piercing whine. "Our environment subsidized travel for the very rich," said Braton. "Some businessman's going to Europe for 10 grand so he can get to a meeting in three hours? People here don't relate to that."

The end of regularly scheduled commercial flights would not mark the last time the supersonic plane was flown, however. Homes for retired, intact Concordes were eventually found in various venues from the

United Kingdom to Germany. The last supersonic flight of a Concorde took place on November 26, 2003, when Concorde Alpha Foxtrot flew to Aerospace Bristol in Bristol, England, where it remains on display. Premier air and space museums vied for and secured surviving Concordes, including the Udvar-Hazy Center of the Smithsonian National Air and Space Museum, outside Washington, D.C.; the Musée de l'Air et de l'Espace at Le Bourget Airport; the National Museum of Flight in East Lothian, Scotland; and the Intrepid Sea, Air, & Space Museum in New York City.

In November 2003, a Concorde mounted on a barge made a stately promenade along the Hudson River to its final destination, the Intrepid Museum. Aviation writer Ron Swanada described the event, which made local and national news, for *Air and Space Magazine*: "The image of the airplane, perched atop the barge like a wounded stork, still proudly holding its head high for all the media's cameras and crowds of onlookers, will be forever in my memory. I can't imagine a more humbling end for an aviation legend."

•

Concorde-related tribulations did not end with the plane's retirement, however. Multiple legal disputes stemming from the crash were pending. Relatives of the victims sought compensation and wanted those responsible to be held accountable. In 2001, Air France reached a $150 million civil settlement with the families of the victims. But French judicial investigators were still at work, compiling 80,000 pages of documents as they considered whether criminal charges were warranted and who would be subject to prosecution. France was then among a handful of nations that routinely built criminal cases targeting individuals and companies involved in major transportation accidents. Among those with reason to be worried were Air France, Continental Airlines, and regulators at France's Direction Générale de l'Aviation Civile, or Directorate General of Civil Aviation (DGAC), the lead government agency responsible for overseeing the safety and security of civil air transport throughout France.

In 2006, executives at Air France and Concorde's manufacturer, Aérospatiale—the precursor to Airbus—were given a preview of what could be in store for them. That year, French prosecutors filed involuntary manslaughter charges against six individuals in connection with the 1992 crash of an Airbus A320 operated by Lignes Aériennes Intérieures, or Air Inter, a domestic subsidiary of Air France that was later absorbed into the parent airline. The six defendants included an Airbus engineer and technical director, the air traffic controller directing the fatal flight, and two employees of Air Inter. Also charged were two former DGAC officials. The charges stemmed from the horrific crash of Flight 148, a night flight from Lyon to Strasbourg. The Airbus A320 slammed into the cloud-shrouded Vosges Mountains near Mont Sainte-Odile on its approach to the Strasbourg Airport on January 20, 1992. Both pilots were killed, along with 82 passengers and three other crew members. Eight passengers and a flight attendant survived the crash.

Investigators said the crash was due to human error. The pilots had mistakenly programmed the flight computer to descend at 3,300 feet a minute instead of a normal 800 feet a minute. The defendants, who could have been sentenced to two years in prison, were charged with causing the deaths of the passengers. French prosecutors cited problems with the design of the cockpit and failure to provide vital safety equipment, which they blamed on a decision by Air Inter to disconnect the ground proximity warning system on all of its planes after experiencing problems with false alarms.

At the time, there was much discussion in the press of the Flight 148 trial as a lead-up to the high-profile trial that seemed likely in the wake of the Concorde crash. It also renewed a heated debate about whether criminal charges should be used as a cudgel to punish government regulators and aviation industry employees. Concerns were raised that airline workers and others with important information to share might not be forthcoming when questioned by investigators following a major crash. "I believe this case has dangerous implications for the criminalization of aviation accidents, which has chilling implications for the sharing of safety data," observed H. Clayton Foushee Jr., a veteran

American air safety investigator who was then scheduled to testify for the defense at the trial.

In May 2006, the case resulting from the crash of Flight 148 went to trial in Colmar, France, and dragged on throughout the summer and into the fall. A verdict was rendered in November, when the presiding judge ruled that Air Inter and Airbus were both liable for compensation to relatives of the victims but at the same time acquitted all six defendants of criminal charges against them. Members of the aviation community cheered the decision, but it drew a mixed response from families of the victims, who had hoped for guilty verdicts, according to Alvaro Rendon, then president of a victims' rights group, Entraide de la Catastrophe des Hauteurs du Saint-Odile. "We are disappointed that these men were acquitted," says Rendon, whose wife was killed in the crash. "But we are also pleased that the truth was acknowledged, and that Airbus and Air France have been held accountable for their responsibility for this tragedy."

The sweeping acquittal of all six defendants in the 2006 Air Inter trial did not dissuade judicial investigators from pursuing criminal charges in connection with the Concorde disaster. In July 2008, a French prosecutor in the Paris suburb of Pontoise announced that the only airline to face criminal charges would be Houston-based Continental Airlines; Air France would bear no blame for the crash. Continental was charged with involuntary manslaughter, as were two of its employees. The Continental workers named were John Taylor—the mechanic whom Continental Airlines said had made the minor repair to the DC-10 that investigators found was the source of the metal strip dropped on the runway—and Stanley Ford, Taylor's supervisor at the Houston maintenance facility where the repair was carried out. Charges of involuntary manslaughter were also filed against French engineers Henri Perrier, the director of the first Concorde program at Aérospatiale, and Jacques Hérubel, who was Concorde's former chief engineer at Aérospatiale. Claude Frantzen, former director of technical services for Concorde at DGAC, was accused of negligence for allegedly ignoring warning signs that had included a string of burst tires over a 15-year period.

After a long and grueling criminal trial that began in May 2010, a lopsided ruling was handed down in December by a judicial panel in Pontoise. The panel of judges decided that none of the highly placed French defendants should be punished. Frantzen was found not guilty, as were Perrier and Hérubel, who were responsible for the testing and certification of Concorde. Perrier and Hérubel's employer, Aérospatiale, was ordered to pay a portion of damages. Blame for the disaster was instead laid at the feet of Continental Airlines and its mechanic, Taylor; both the company and Taylor were found guilty of involuntary manslaughter, although Taylor's supervisor, Ford, was acquitted. Taylor was sentenced by the French court to a 15-month suspended sentence for involuntary manslaughter. Continental Airlines was held liable for damages of €1 million ($1.3 million) to Air France as compensation for damage to Air France's image.

At the time, the lawyer for Continental Airlines, Olivier Metzner, attacked the ruling, saying it was "incomprehensible to put everything on the Americans' shoulders." Continental, which in 2010 had been absorbed into United Airlines, found an unexpected ally in Roland Rappaport, the lawyer for the family of the late Air France captain Christian Marty, who maintained that the court had overlooked the role played by French engineers and safety regulators. "I don't understand the difference of treatment between the American transporter and those who were involved on the French side," said Rappaport.

The legal war was not over, however. Appeals were soon filed, and the battle front shifted to a court in Versailles. In November 2012, an appeals court judge overturned the guilty verdicts for both Continental Airlines and its mechanic, absolving both the company and Taylor of involuntary manslaughter. In her ruling, appeals court Judge Michele Luga said that any mistakes made by Continental and Taylor didn't make them criminally responsible for the deaths. "He could never have imagined a scenario where this simple titanium blade could cause such a disaster," Luga said in court of Taylor. Luga did, however, uphold the lower court's finding that the dropped metal strip was the precipitating cause of the crash. Continental Airlines therefore would still be liable for damages to be paid to Air France.

Luga also went on to make a point of scolding French civil aviation authorities for their lax performance, citing "25 years of operation . . . littered with numerous cases of tire damage following more or less serious incidents." Confronted with the pattern of failures, Luga said, regulators should have stepped in and suspended Concorde's airworthiness certificate. Luga's rulings and comments were met with mixed reactions from an attorney representing families of the victims. "The court says the plane shouldn't have flown," says Stéphane Gicquel, a victims' rights advocate and spokesman for a group made up of families of the Concorde crash victims. "It did fly, but no conclusion is drawn." The outcome, he says, left his clients with unanswered questions and a feeling of powerlessness.

There would be one more twist in the long, torturous legal saga that began in the courtrooms of France. In December 2013, Continental mechanic Taylor filed suit in Chicago, in Cook County, against United Airlines, United Continental Holdings (UCHI), and Air France–KLM. Taylor's suit claimed that his former employer, which had been purchased by United Airlines, had used Taylor as a scapegoat. "He truly is the last victim of the Concorde crash," his Houston-based attorney, Paul LaValle, said. The French prosecutor who filed the charges against Taylor was overzealous, said Gary D. McCallister, the Chicago-based attorney who helped Taylor file his complaint. "To claim one man is responsible for 113 deaths was a massive allegation, and one that was simply untrue." Both LaValle and McCallister said that they were prepared to argue in court that Taylor did not perform the work on the DC-10 in question. "One thing we were confident of is that John Taylor did not do that repair," said McCallister. At the time that Taylor was working for Continental, all mechanics were required to enter any work performed in a logbook, which was signed off by a supervisor. "That never happened," says LaValle. "Taylor never worked on it [the DC-10] or filled in the logbook. There's no documentation that the work was ever done by him."

Even though both Taylor and Continental were eventually absolved of charges of involuntary manslaughter, the entire ordeal was personally devastating for Taylor, according to LaValle. "He got divorced, became alcoholic and suicidal, and experienced all the emotions anybody would go through in his situation." Before the 2010 criminal trial in France, Taylor worried that he might be seized and extradited by bounty hunters, according to LaValle. In 2010, Continental instructed Taylor not to go to France to testify in the criminal trial stemming from the crash. As a result, he was unable to defend himself in court; Continental's attorneys maintained that Taylor had performed the repair but that neither he nor the airline were responsible for the crash.

The 15-page lawsuit filed by Taylor in 2013 was not able to proceed, however. The judge who reviewed the lawsuit dismissed it, saying it should not have been filed in Cook County because the crash had taken place in France. LaValle says that Taylor, still hoping to clear his name, hired a French attorney to take up his complaint, but the attorney advised LaValle and Taylor that no French court would be receptive to his claim, and that he should not waste more of his time and money pursuing it.

•

As time passed, the legacy of Concorde would be dominated not by the emotional scars left by the crash, but by the plaudits of aviation enthusiasts and fond memories of its passengers and fans. In 2017, the BBC published a series of personal accounts written by people whose lives and relationships were intertwined with Concorde's role as a magical carpet that whisked them out of the routine of their daily lives into another dimension defined by luxury and adventure. One contributor, Trish Dainton of London, recalls the trip she took with her late husband, Steve Dainton, who later died of Huntington's disease. The couple had saved to make the flight to celebrate their 10th wedding anniversary. They met the captain and took photos with the crew. "I will always cherish having been able to share living the dream with Steve by my side," writes Dainton. "I hope Concorde takes to the skies again one day to allow others to dream and to fulfill their dreams like we did."

In another account, amateur photographer Jetinder Sira recalls her youthful encounters with Concorde. "The end of the runways at Heathrow all have public roads and footpaths near them," she writes, and she found herself drawn to these busy outskirts during the summer of 2003. She brought with her a camera and stool on which to perch to get optimal shots of the sleek and powerful plane as it flew heavenward:

> The idea was to stand on the stool near the end of the runway, peek over the perimeter fence and look down the runway. When I saw Concorde coming I got my camera and started taking photos. As she very rapidly flew over the fence and over my head, she formed a shockwave and this shockwave knocked me off my feet and off my stool. It was an awesome experience. It wasn't a one off, it happened on every single takeoff. No other plane, not even the "overweight" Boeing 747, could ever knock me off my feet, but Concorde did.

Concorde as a static museum exhibit will remain an inspiring sight. But it will never again soar 11 miles above the earth or knock fascinated spectators off their feet. To fulfill its promise, Concorde requires an heir in the form of an aircraft that will be the product of an even more innovative design, one that would make possible a second generation of supersonic planes. If the future of civil transport is to include another fantastic jetliner that can fly faster than a bullet, 21st-century engineers will have to achieve what their 20th-century counterparts could not: design an aircraft that is not only incredibly swift but also efficient, affordable, reasonably quiet, and capable of inspiring a new generation of aviation dreamers.

13

A Supersonic Future

Concorde was always meant to have a successor, what its creators called a B model, to be launched in 1982. But reality had other plans, and the world's only supersonic passenger plane was permanently retired in 2003 without a replacement. More recently, however, talk of a suitable heir for Concorde has gone beyond "what if" to "what's next."

In one version of what could come to pass, the supersonic plane of tomorrow will be a sleek, windowless business jet flown by pampered VIPs lounging in club seats as they zoom from San Francisco to Tokyo to close a multimillion-dollar deal. In the other, more democratic version of the near supersonic future, Concorde's heir will be a midsize passenger plane with half as many seats as Concorde but with tickets selling for less than half the price, making supersonic travel an affordable option for families willing to pay business-class prices for a swifter flight from New York to Sydney.

Whether Concorde's inheritor will take the form of a corporate jet flown by the few or a passenger plane with room for the average traveler, any second-generation supersonic aircraft designed for the civil aviation market will have to be much quieter than its Anglo-French predecessor. And it will almost certainly be designed and built not in Europe but in the United States. Multiple American aerospace research and development companies—ranging from big and prestigious to small and plucky—are currently engaged in a competition to build and fly the first supersonic commercial plane designed, tested, and assembled at home. The ongoing race is taking place in tandem with research underway at the National Aeronautics and Space Administration, which is investing its considerable engineering talent and millions of dollars in a quest to take the boom out of sonic booms and open the way for overland routes to be used by the next generation of quieter supersonic jets.

If the iconic Concorde is to have a suitable heir while the 21st century is still young, it could be a plane named, ironically, Boom—the brainchild of an engineer from the Midwest named Blake Scholl, the chief executive officer and cofounder of Boom Technology. Depending upon how pessimistic or optimistic your thinking is on the prospects of *any* supersonic plane being built in the next decade (ranging from "it won't happen" to "get ready for it"), such a plane might never make it to market, or it could be flight-tested and sold to major carriers around the world by 2023.

That's the date set by Scholl, a pilot with an entrepreneurial streak who thinks that supersonic flight should make the world a smaller place for everyone, not just those able to pay for a pricey seat in a chartered jet. When Scholl was a boy, his family made regular visits to spend time with his grandfather, who lived just 90 minutes away by car. "I saw him every weekend and we got super close," says Scholl. "I wouldn't be doing what I'm doing if it wasn't for his life lessons." Now Scholl and his family live in Denver, where his startup is developing a supersonic passenger jet to sell to airlines both domestic and foreign. "Fast-forward 30 years and now I have kids, and their grandpa is in Hong Kong. That's 18 hours away from my home in Denver, and my kids are only going to see him once a year."

Scholl says the answer to the problem of distance separating his family, and countless others, is to radically reduce intercontinental flying times. Doing so won't be easy to achieve for a number of reasons, but he says a supersonic revival is long overdue. "We're 60 years into the jet age and we're literally still in the jet age," he says. "We're about to have self-driving cars, but we're still flying around at 1960s speeds. . . .

We had a supersonic plane that was twice as fast, but we never took it mainstream."

Scholl and his design team at Boom Technology are the architects of the XB-1, nicknamed Baby Boom, a two-seat demonstrator aircraft and pathfinder to what will be a 45-to-55-seat aircraft designed to fly at Mach 2.2. Boom says their supersonic plane will be lighter and faster than Concorde. Scholl also insists it will be cheaper to fly. At least that's his goal. "This isn't a private jet," says Scholl, who cofounded the company in 2014 with chief engineer Joe Wilding and chief technology officer Josh Krall. "We want to build something that we can see our friends and family flying on. We're starting with business-class prices because that's what we have technology for. But our line of sight is we want to make the fastest ticket the cheapest ticket."

If the path to affordable tickets on supersonic planes seems far off and chock-full of pitfalls along the way, that's because it is—a reality Scholl acknowledges. "When I started looking into this, nothing much was going on in the supersonic market," Scholl told BBC News when he attended the Dubai Airshow in November 2017 on a mission to cultivate superrich investors. "There was some sci-fi stuff. I thought it was probably impossible. But after researching the field, I started to think—it's possible, but difficult. If you have enough courage and you get the right people together, you can do it, I thought."

Scholl, who studied computer science at Carnegie Mellon and became an engineer turned Amazon executive before founding his own mobile payments firm, Kima Labs, never flew on Concorde before its retirement in 2003. But he was inspired by what its designers achieved in the precomputerized era of slide rules and cheap oil and hopes to engineer a plane that will be lighter and more fuel-efficient. "The fuel economy [of Concorde] was awful," Scholl says, "and that drove prices up right there." The fledgling company aims to fly high and fast: if Boom delivers on their promises, their supersonic plane will be 10 percent speedier than Concorde and able to make the transatlantic flight from New York to London in three hours and 15 minutes. They say they'll charge for the flight only the cost of an ordinary roundtrip business-class ticket, about $5,000—far less than the $13,500 cost of a roundtrip ticket on Concorde in 2003.

In November 2016, Boom unveiled a static mockup of what will be the prototype aircraft known as Baby Boom. The prototype design is complete, but the two-seat Baby Boom has not yet made its debut. The company says that it is on track to be manufactured in North Carolina, then brought to Denver's Centennial Airport in late 2018 or early 2019 for initial high-speed test flights before undergoing supersonic flights in the test corridor out of Edwards Air Force Base in California. At least that's the plan. In the meantime, Boom already has used a miniature model to conduct successful subsonic wind tunnel tests at Wichita State University in Kansas.

Boom will need more investors, but it has not lacked for entrepreneurial hustle. The startup reports that it has presold 76 of its planes, 10 of them to Richard Branson of Virgin Galactic, the launch customer for the XB-1. Branson famously tried more than once to buy a Concorde from British Airways but was rebuffed. Now his Spaceship Company, the manufacturing arm of Virgin Galactic, is backing Boom with engineering and manufacturing expertise and services. In December 2017, Japan Airlines announced that it was investing $10 million in the startup, whose planes are expected to sell for $200 million each. Japan Airlines also has an option to buy 20 Boom planes and has agreed to help the young company with its design and passenger experience.

Scholl points out that the technology for Boom has existed for more than 50 years, while admitting that the original drawbacks that stymied Concorde—sonic booms and fuel inefficiency—remain to be conquered. How does Boom intend to accomplish what its predecessor, backed by the combined resources of two rich nations, could not? Boom is counting on a new, more aerodynamic shape with wing extensions designed to increase lift and a refined delta wing that will reduce supersonic drag and quiet supersonic booms. New carbon-fiber composites will make its jet lighter than Concorde, too, says Boom. That will help cut fuel costs, but achieving commercial success is still a long shot. "Supersonic business jets I still regard as inevitable," says Richard

Aboulafia, vice president of analysis at aerospace and defense analysis firm Teal Group. "Commercial transports? Boy, it could go either way."

An American-made SST that is quieter than Concorde might have a chance, however, and getting a regulatory reprieve from the 1973 ban on supersonic flights over the United States will be vital to the plane's long-term economic survival. In the meantime, Boom is talking about flying over water on routes that could include New York to London, San Francisco to Tokyo, and Los Angeles to Sydney, while not ruling out the possibility of overland flight corridors.

Even if the plane is 25 to 30 percent quieter, as Boom forecasts, regulatory barriers will have to be adjusted to clear the way for supersonic travel overland in the United States. "Supposedly that's about noise," Scholl says. "I think it had to do with protecting Boeing from Concorde"—a reference to giant Boeing Company's early efforts to design and test a supersonic passenger plane by 1974 and American governmental resistance to competition from Concorde. But just as loud as Concorde's boom was the outcry from members of the American public, who didn't want their windows rattled and their nerves shattered by the daily sonic booms overland flights would produce.

Will a 25 to 30 percent noise reduction be enough to make sonic booms bearable? In 1964, residents of Oklahoma City were subjected to a daily barrage of eight sonic booms a day from flyovers by F-104 Starfighters as part of Operation Bongo, a six-month-long experiment conducted by the Federal Aeronautics Administration. The federal agency was soon overwhelmed with noise complaints and thousands of damage claims, many for cracked sheetrock and broken windows. "The first time that it happened, I was visiting my mother," recalls Bridget Meadows, 65. "She lives in Oklahoma City, and my children were playing in her backyard at the time. The first boom that I ever heard—I didn't know what it was at the time—it literally shook the house. The dishes rattled. It was kind of like what I always imagined an earthquake would be. The kids ran in crying."

Supersonic planes don't just boom at high altitudes; their engines are also noisier during takeoff and climb-out. Without a solution to the problem of engine noise, which depends in part upon as-yet unrealized engineering improvements required to make supersonic engines both less polluting and quieter, talk of overland supersonic flight remains mere conjecture.

There have been signs, however, that the mood in Washington, D.C., is shifting and regulatory change is in the wind. In July 2017, lawmakers in Congress directed the Federal Aviation Administration to revisit the subject of what constitutes acceptable noise levels generated by sonic booms and to report back. The stated aim was to clear the way for overland supersonic flights, assuming that American-made planes under development would be low-boom and therefore quieter than Concorde. As a result, the FAA began revisiting regulations governing supersonic aircraft and studying advancements in aircraft design that could reduce noise pollution, raising the possibility of exemptions from takeoff noise requirements for supersonic jets, as well as a new supersonic noise standard that could set a different bar for what constitutes an acceptable sonic boom.

That push by Congress coincided with funding for NASA to collect data on what might be considered acceptable noise levels for sonic booms over urban areas. That question won't be easy to answer, given variables both scientific and subjective. A sonic boom is not a one-time noise event but a continuous noise generated all along an aircraft's route as it flies at supersonic speed. A plane flying supersonic at 50,000 feet can produce a sonic boom cone about 50 miles wide, although parts of the sonic boom "carpet" are typically weaker than others. A Concorde traveling at 50,000 feet sometimes exceeded noise levels designated as painful to the human ear and roughly equivalent to highly amplified music or the sound of an automatic riveting machine. While the sonic-boom shock wave from a Concorde flying at Mach 1.93 at 48,000 feet raised the pressure on the human ear only 1.94 pounds per square foot above normal atmospheric pressure, that increase happened so quickly—in about half a second—that it created a perceived noise of 105 decibels on the A scale, which takes into account the sensitivity of the human ear to different frequencies of sound. The typical Concorde sonic boom noise fell somewhere between 90 A-scale decibels—the noise of a lawnmower—and the 110-decibel sound of a jackhammer or power saw. (A nearby jet takeoff produces a noise level of 130 decibels.)

Addressing the noise pollution problem is key to any hope for a supersonic revival—a reality not lost on NASA, where the pursuit of so-called low-boom technology is underway as part of a Quiet Supersonic Technology program. Its goal is to develop the technology necessary to replace noisy sonic booms with a softer thump sometimes compared to a heartbeat. With its $19 billion annual budget and 19,000 in-house employees, NASA's commitment to developing low-boom technology could be a game changer. It has partnered with Lockheed Martin and its famed Skunk Works engineers in Palmdale, California, to design and test a quiet supersonic aircraft as part of a $20 million contract awarded in February 2016 to Lockheed and subcontractors GE Aviation of Cincinnati and Tri Models Inc. of Huntington Beach, California. The experimental aircraft under development has been dubbed the LBFD X-plane, short for Low Boom Flight Demonstrator X-plane.

The X-plane designation makes it a member of the distinguished club of early experimental planes—primarily military aircraft—developed by NASA and Lockheed Martin and designed at Skunk Works under the direction of Clarence "Kelly" Johnson, one of the most highly regarded aircraft designers in American aviation history. The supersonic X-plane design will be the working blueprint for a demonstrator aircraft called the QueSST that could be flight-tested by 2020 or 2021. The 90-foot-long QueSST will simulate the sonic boom of a 100-passenger supersonic plane. "We think that this [the X-plane and its spinoffs] will be the next generation of commercial transport by air," says John Carter, manager of NASA's Quiet Supersonic Technology preliminary design team, based at Armstrong Flight Research Center in California. "We think it is going to be a new industry, and we want to make sure U.S. companies are the first to market with supersonic commercial vehicles."

NASA's ambitious goal of overseeing production of a low-boom X-plane has an active corollary in the private sector, where two aerospace companies are working on designs for supersonic business jets. Both Boston-based Spike Aerospace and Reno-based Aerion Corp. are engineering supersonic business jets: the S-512 by Spike and the AS2 by Aerion, which is getting help from Lockheed Martin and Airbus, Europe's leading aerospace manufacturer. Aerion is touting its AS2 as an 8- to 12-passenger plane with a low-drag "laminar flow" wing that will fly at Mach 1.4. Not to be outdone, Spike is developing its own speed demon; the 12- to 18-passenger S-512 business jet will zoom along at a maximum speed of Mach 1.6, making it possible to fly from Dubai to Paris in 3.1 hours "without creating a disturbing sonic boom."

It will be quieter than Concorde, producing what its designers say will sound like a "soft clap" to anyone at ground level. Spike aims to sell its jet to the public as environmentally friendly, with "low impact" on migratory birds and marine life, according to the company's promotional materials. Instead of windows, the S-512 will feature cabin-length panoramic views of the outside world on high-definition screens spanning each side. Spike touts them as "revolutionary displays [that] can show any view you wish, whether it's your real-time aircraft surroundings, your favorite movie or a work presentation—all with a simple touch of your mobile device." Spike's plans call for it to test-fly a supersonic demonstrator in 2019 and deliver its aircraft to customers by 2023.

While both of these supersonic business jets are snazzy and sleek, they are far from revolutionary in terms of their impact on civil aviation, since business jets make up less than 4 percent of the civil aviation market. More promising in that regard is NASA's QueSST, the pathfinder for a plane that could carry 100 passengers.

All the participants in the race to design a new, quieter supersonic jet are facing the same challenges as Concorde's designers, but they have more tools at their disposal to solve these problems. In the 1960s, Concorde engineers did their calculations with slide rules, made their drawings on paper, and tested their designs on physical models. Current SST designers have the advantage of computer-assisted design systems and simulation software for testing. The most obvious differences between these newer SST designs and the Concorde are their radically different aerodynamic configurations and smaller size, both of which were developed to reduce the sonic boom problem.

The NASA X-plane, for example, relies on a needle-nosed fuselage with a narrow delta wing to produce a softer thumping sound rather than the sharp, double-cracking sonic boom made by Concorde. Its futuristic look includes a long, slender fuselage, a highly swept-back delta wing, and multiple control surfaces that can tailor the distribution of pressure and lift. "You try to minimize the strength of the shock, so you have a very sharp nose on the aircraft, but you have to extend it a long way past the fuselage," explains Thomas Corke, professor of engineering at the University of Notre Dame and director of the university's Institute for Flow Physics and Control. "The idea is that there is no way to avoid a shock wave supersonically. The [X-plane] design doesn't eliminate shock; it just minimizes it, so what's [heard] on the ground is almost imperceptible."

The smaller size will also reduce engine noise, since less far less thrust will be needed. However, any plane large enough to carry more than a handful of passengers must be equipped with engines that are both powerful and quiet. Therein lies the crux of the problem confronting Boom and any other would-be developer that wants to reach a larger market. The Boom designers are depending upon engine manufacturers such as General Electric to come up with a revolutionary engine design flexible enough to span the demands of supersonic, transonic, and supersonic flight.

But in addition to being far less noisy than Concorde engines, the Boom power plants will have to be more fuel-efficient and ecologically sound. Can that be done? It's a tall order, especially considering that any manufacturer must be willing to invest millions of dollars while knowing the demand for their engines will be limited by market competition between new supersonic planes and the dominant wide-body planes. Without new engines that check all the boxes for genuinely cleaner and more fuel-efficient power, Boom will face the same economic hurdles that Concorde failed to clear.

NASA, in partnership with ambitious American companies, may have the means and talent to retune sonic booms from a thunderclap to a thump. But even that much noise, multiplied by regular daily flights, may prove unpopular. Reduced travel times are not a guaranteed way to win over people living near airports and underneath supersonic flight corridors. Many questions remain as well about whether NASA can deliver on its pledge to develop engines that will be not simply more efficient but also cleaner.

All these technical details are complicated by the larger cultural discussion about whether we actually need a faster future. So far, most technological improvements aimed at saving time have only accelerated the pace of modern life, packing every 24-hour cycle with more tasks that need to be accomplished more quickly. While dramatically reduced flight times could prove very appealing, no one really knows if the ability to fly halfway around the world in three hours will bring about a revolution in travel and create a more connected global community, or if it will further separate the minority who can afford a streamlined, stress-free travel experience from the majority who cannot.

In the meantime, the prospect of new supersonic planes is already generating more talk of the next bold technological step forward: planes that fly at *hyper*sonic speed. Industry leader Boeing is at work on a second-generation spy plane that could travel at five times the speed of sound—making it capable of outrunning a cruise missile and disappearing in the blink of an eye. Nor does Boeing shy away from blue-sky talk of hypersonic passenger planes that would fly at Mach 5 and take a mere two hours to shuttle between New York and Shanghai. "I think in the next decade or two, you're going to see them become a reality," says Boeing chairman and CEO Dennis Muilenburg.

•

While the future beckons, the past lingers, leaving an indelible imprint. Not just one but two impressive monuments have been erected in

France to honor those who died in the Concorde crash of 2000. One is located at the crash site in Gonesse, where the Hotelissimo once stood between two busy roadways. The other is on a quiet, tree-lined parcel of land near Charles de Gaulle Airport. The Gonesse monument, paid for with funds raised by locals, is an impressive 7-meter- (23-ft-) high panel of transparent glass pierced by a splinter of steel resembling an aircraft wing. On July 25, 2010, the 10th anniversary of the crash, about 70 Gonesse residents and relatives of the victims gathered there for a Sunday-morning memorial service, including many who had traveled from Germany. The roar of planes taking off from nearby Charles de Gaulle could be heard during the ceremony's long moment of silence; Air France officials gave red and white roses to the relatives to place at the foot of the monument.

After the ceremony, many wandered the field behind the monument, where Concorde had fallen to earth. Among those in attendance was Claudine Le Gouadec, the sister of Virginie Le Gouadec, chief flight attendant on Flight 4590. "I still have trouble believing that she is gone," she said. "It still seems abstract to me, the loss. For me she still exists, but I don't see her." Also present were many who live in the small city that is neighbor to the busy airport, including some who expressed lingering fears of another deadly crash. "I don't want to call it an obsession, because we are so used to hearing the noise" of aircraft, said Gonesse resident Claude Philippe. "But sometimes when there is a weird sound, we say to ourselves, 'Here it is, this one is for us.'"

The other monument is appropriately located quite near where the accident began and was designed with careful attention to the events of July 25, 2000. Adjacent to a perimeter road running south of Charles de Gaulle, it is a formal, somber parklike setting defined by 113 low bushes arranged in the shape of a Concorde. The imaginary plane's longitudinal axis is aligned with runway 26R, where Concorde F-BTSC exploded into flames just before takeoff. A pathway leading to the monument is bordered by 113 plaques inscribed with the names of the dead. At the end of the pathway is a vertical sculpture of metal and rock: an inverted steel triangle, representing the doomed plane positioned nose downward, is pressed between two upthrust, jagged slabs of stone. An inscription in French, German, and English pays homage to the dead and explains that the sculpture is meant to both "conjure up the tremendous force of nature" and evoke nature as a source of "strength and support." It concludes by noting that the sculpture, like Concorde itself, "symbolizes both the achievements and limitations of technical feasibility."

•

Concorde's evolving legacy surely will be enhanced if future events cast it in the role of progenitor to subsequent generations of supersonic and hypersonic planes. Alternately, it could be consigned to a very different place in the pantheon of innovative aircraft, constituting a fascinating and bold aberration that nonetheless led to an evolutionary dead end. Over time, it has become clear that it was not Concorde's status as a supersonic plane that caused the catastrophe. Instead, the factors that led to the crash were a more familiar combination of unforeseen, random events, in conjunction with institutional reluctance on the part of both regulators and air carriers to address known safety concerns, meeting them with short-term fixes instead of more costly remedies that might have prevented the crash—an unfortunate alignment of circumstance and error similar to that revealed by most air crash investigations. The events of July 25, 2000, constitute a singularly dark episode in the otherwise uplifting narrative of an iconic aircraft built via international cooperation. Concorde's mesmerizing image still endures, both inspiring and daring would-be aviation pioneers to fulfill its glittering promise.

Sources

1. The White Bird

Information in this chapter about weather conditions and their effects on passengers making connecting flights at Charles de Gaulle on the morning of July 25, 2000, is drawn from Jason Burke, "Death of a Dream," *Observer*, July 30, 2000.

Other weather-related details—including low-pressure systems in Europe, local temperatures at the airport, and local visibility on the day of the crash—are drawn from Ministère de l'Equipement, des Transports, et du Logement, Bureau d'Enquêtes et d'Analyses pour la Securité de l'Aviation Civile, France, "Meteorological Conditions" and "Situation at the Aerodrome," in *Accident on July 25, 2000, at La Pattie d'Oie in Gonesse (95) to the Concorde Registered F-BTSC Operated by Air France*, Report Translation f-sc000725a, English ed. (January 2002), 36, https://goo.gl/TwXZ6s. (Hereafter *BEA Report*.)

Quotes and observations by air traffic controller Gilles Logelin, as well as information on the volume of activity at the airport, the types of planes that land there, the duties of air traffic controllers, and the number of movements at the airport during a 24-hour period, are drawn from the transcript of an interview conducted by Caroline Buckley with Logelin, April 8, 2014, Aéroport de Paris–Le Bourget, France, for *Air Crash Investigation*, episode 7, Concorde AF4590. (Hereafter *Air Crash Investigation*.)

Gilles Logelin's excerpted radio communications are drawn from the transcript of the cockpit voice recorder, Appendix 2, "CVR Transcript," *BEA Report*.

Architectural information and other details about the southern control tower structure at Charles de Gaulle Airport are drawn from "Roissy Charles de Gaulle Airport: Satellite A and New Control Tower," *International Architecture Yearbook: Millennium Edition*, no. 6 (Victoria, Australia: Images Publishing, 2000), 226.

Information about the length of the runways at Charles de Gaulle Airport, the location of its runways, and which runways were in use in 2000 is drawn from "Aerodrome Information," *BEA Report*, 39–40.

The reference to 148 movements per hour at Charles de Gaulle Airport is taken from Pierre Leroyer, *Airside Study of Charles de Gaulle Airport* (Boston: Massachusetts Institute of Technology, December 2004), 7, https://goo.gl/EL8gNi.

Details about the number of taxiways, runways, and terminals and annual passenger traffic at Charles de Gaulle Airport are drawn from "Paris CDG Is Second Busiest in Europe," *Airline Network News and Analysis*, June 9, 2017, https://goo.gl/EYPwk8.

The anecdote about the launch, landing, and destruction of the world's first hydrogen gas balloon in Gonesse in August 1783 is drawn from Glen Bledsoe and Karen E. Bledsoe, *Ballooning Adventures* (North Mankato, MN: Capstone Books, 2001), 16–17.

The scheduled 3:25 p.m. departure time of Flight 4590 is drawn from media reports, including Anne Swardson and Don Phillips, "Crash Probe Centers on Jet Engine," *Washington Post*, July 27, 2000. (Hereafter Swardson and Phillips, "Crash Probe.")

Information on the number of workers employed at Charles de Gaulle Airport is drawn from "40-Year Milestone for Paris–Charles de Gaulle Airport," *Airport Focus International*, April 24, 2014, https://goo.gl/9FA3Mx.

Details about the hotels in the commune of Roissy, its atmosphere and character, and the aviation monuments and art installations there are drawn from Bob Lyons, "Roissy en France: The Village That Hosts the World," *The Good Life: France*, n.d., https://goo.gl/T5BtZf.

Details and descriptive quotes about the cruise ship MS *Deutschland* are drawn from Peter Hughes, "Cruises: On Board MS *Deutschland*," *Telegraph*, November 11, 2008.

Details about Concorde F-BTSC's date of entry into service, date of first flight, number of hours flown, and number of landings are drawn from Christopher Orlebar, *The Concorde Story*, 7th ed. (Oxford: Osprey Publishing, 2011), 230. (Hereafter Orlebar, *The Concorde Story*, 7th ed.)

Dates and other details about the maintenance of Concorde F-BTSC are drawn from "Maintenance," *BEA Report*, 21.

Quotes and observations by Concorde pilot and captain Jean-Louis Chatelain are drawn from the transcript of an interview conducted by Caroline Buckley with him, April 7, 2014, Nice, France, for *Air Crash Investigation*.

The estimate of annual hours flown by a typical Boeing 747 is based on the number of 747s that have exceeded projected 20-year lifespans flying an estimated 60,000 hours over 20 years or 3,000 hours per year, as reported by Matthew L. Wald, "747 Fleet's Age at Issue During Flight 800 Hearing," *New York Times*, December 12, 1997.

Biographical details and information about pilot Christian Marty's accomplishments as a windsurfer and sportsman and details about his record-breaking 1982 windsurfing journey are drawn from "Pilot Was One of Few to Have Windsurfed Across the Atlantic," *Independent*, July 26, 2000; see also Laurie Nadel, "A Tribute to Christian Marty," *American Windsurfer* 7, no. 5, https://goo.gl/jSDAN2.

Information about the July 24, 2000, report of cracks found in the wings of British Airways planes is drawn from Matt Born, "Concorde Grounded after Cracks Are Found in Wings," *Telegraph*, July 24, 2000.

Biographical details, names, and ages of Concorde passengers Kurt Kahle, his wife, Marion Kahle, and his son Michael Kahle are drawn from Tony Paterson and John Hooper, "Germany Struggles with Its Grief," *Guardian*, July 26, 2000.

Biographical information about the life and careers of Irene Vogt-Götz and her husband, Christian Götz, and their battles with cancer are drawn from Steve Boggan, "A Tragic Tale Emerges from the Wreckage of Concorde," *Independent*, August 3, 2000; see also "Families Wiped Out in Crash," BBC News, July 31, 2000.

Biographical details, names, and ages of Concorde passengers and married couples Rolf and Doris Maldry, Andreas and Maria Schranner, Andrea and Christian Eich, and their children, Maximilian Eich and Katharina Eich, are drawn from Lucian Kim and Steve Boggan, "Germans Saved for Years to Go on Supersonic Trip," *Independent*, July 27, 2000.

Biographical details, names, and ages of Concorde passenger Klaus Frentzem and his wife, Margaret Frentzem, are based on reporting by Hannah Cleaver, "Three Generations Died on Concorde," *Telegraph*, July 28, 2000. Details about Klaus Frentzem showing off his collection of Concorde memorabilia, including toy airplanes, that he had brought with him for the trip are drawn from Burke, "Death of a Dream."

The reference to a July 1985 charter flight for 60 children who won prizes that earned them a free trip to Legoland Billund, Denmark, is drawn from Orlebar, *The Concorde Story: 21 Years in Service* (London: Osprey Publishing, 2004), 116. (Hereafter Orlebar, *The Concorde Story: 21 Years in Service*.)

The excerpted quote from travel writer Graham Boynton is drawn from his "A Flight of Fancy That Made Your Spirit Soar," *Telegraph*, July 26, 2000.

The detail about the intentional omission of row number 13 in Concorde passenger cabins is drawn from Nathan Roemer, "Welcome to Concorde," *Travel Scholar*, 2003, https://goo.gl/YsDH7j.

Information about the Concorde's Rolls-Royce/SNECMA Olympus 593 engines and the plane's fuel consumption per gallon per passenger can be found in Gordon Roxburgh, "Concorde Technical Specs: Powerplant," *Concorde SST*, https://goo.gl/aroJGM.

Quotes and observations by Bob MacIntosh in this chapter are drawn from the transcript of an interview conducted with him by Sonya Buyting, March 27, 2014, Hamilton, Ontario, for *Air Crash Investigation*.

2. "You Have Flames behind You"

Unless otherwise indicated, all quotes in this chapter by pilots Christian Marty, Jean Marcot, and Gilles Jardinaud, as well as quotes by unidentified flight attendants, a dispatcher, and other unidentified persons communicating via radio, are drawn from Appendix 2, "CVR Transcript," *BEA Report*, n.p.

Details about preflight preparations for Flight 4590, Captain Marty's decision to replace the defective thrust reverser in the number 2 engine, and repairs undertaken while the plane sat parked at the gate are drawn from "Flight Preparation" and "The Flight until Engine Power Up," *BEA Report*, 158; see also Swardson and Phillips, "Crash Probe"; and Burke, "Death of a Dream."

Times indicated for events prior to takeoff, including the completion of the repair to the thrust reverser, are drawn primarily from Appendix 2, "CVR Transcript," *BEA Report*, n.p.; see also Orlebar, *The Concorde Story*, 7th ed., 149; and Swardson and Phillips, "Crash Probe."

Information about the late arrival of the connecting flight carrying Concorde passengers from Germany and the description of the carefree mood of passengers as they waited to board are drawn from Burke, "Death of a Dream."

The V speeds set by the flight crew before takeoff can be found in "Factual Information," *BEA Report*, 17.

Details about the professional background and technical qualifications of the three-person Concorde flight crew are drawn from "Personnel Information," *BEA Report*, 18–20.

The weight of the luggage on board, the plane's takeoff weight, the total fuel weight, and the engine startup sequence are drawn from "The Flight until Engine Power Up," *BEA Report*, 159.

The source used for the average takeoff speed of Concorde is Gordon Roxburgh, "Concorde Technical Specs," *Concorde SST*, https://goo.gl/wW19F7.

Information about average surface wind speeds in the area of the airport and runways 26R and 26L on the afternoon of July 25, 2000, is drawn from "Situation at the Aerodrome," *BEA Report*, 36.

The time of day that air traffic controller Gilles Logelin reported seeing flames behind the plane, and the distance the plane had traveled along the runway at that point, is referenced in "The Flight Up Until the Loss of Thrust on Engine 1," *BEA Report*, 161.

Times that Concorde F-BTSC's various alarms were heard are noted in "Transcript of the Recording," *BEA Report*, 48, and Appendix 3, "Analysis of the Alarms and Noises Recorded on the CVR," *BEA Report*, 1.

The plane's airspeed of 200 knots at 4:43:27 p.m. and also later at 4:43:49 p.m. is found in "The Flight Up Until the Loss of Thrust on Engine 1," *BEA Report*, 163, 164.

The distance of 9,500 meters (9.5 km, 5.9 mi) from the threshold of runway 26R to the crash site in Gonesse can be found in "The Accident Site," *BEA Report*, 67.

Quotes by American pilot Sid Hare, airport employee Julian Pyke, driver Frederic Savery, and witness Samir Hossein are drawn from Patrick Bishop and Harry de Quetteville, "113 Killed in Concorde Crash," *Telegraph*, July 26, 2000.

The reference to the site of the crash as located at the intersection of the N17 and D902 roadways is drawn from "The Accident Site," *BEA Report*, 67.

The reference to the video of the Concorde on its flight path that was taken from the cab of a moving truck, driven by an anonymous Spanish couple as they passed Charles de Gaulle, can be found in multiple media reports, including "Concorde Crash," *Guardian*, July 26, 2000; and Amanda Killelea, "British Concorde Plane Crash Survivor Speaks about How the Tragedy Changed Her Life 15 Years On," *Mirror*, July 24, 2015.

Quotes by Cambridge University student Alice Brooking and the description of her escape from the hotel are drawn from "Cambridge Student Leapt from Window to Escape Burning Building," *Independent*, July 26, 2000.

Biographical details, names, and ages of Hotelissimo manager Michèle Fricheteau and the four hotel employees killed in the crash are drawn from "Forgotten Victim," *Irish Times*, August 12, 2000; see also Sean O'Neill, "Don't Forget My Staff Who Died, Says Hotel Owner," *Telegraph*, July 28, 2000.

3. A Lunar Landscape

Times assigned to the start of the fire that began during Flight 4590's takeoff, and the description of the initial physical location as being beneath the left side of the plane's wing, are drawn from "Fire," *BEA Report*, 87.

Unless otherwise specified, times cited regarding events before, during, and after takeoff are drawn from Appendix 2, "CVR Transcript," *BEA Report*, n.p. The exact time of rotation is based on information in "History of the Flight," *BEA Report*, 17.

The quote by Patrick Tesse, manager and owner of Hôtel Les Relais Bleus (adjacent to the destroyed Hotelissimo), in Gonesse, about his reaction to the impending crash is drawn from Harvey de Quetteville and Michael Smith, "It Was Like an Atom Bomb in the Sky," *Telegraph*, July 26, 2000; as well as John Henley, "Concorde Families Await Justice," *Guardian*, July 25, 2002.

Information about the bodies of crash victims being taken to a makeshift morgue in Gonesse are drawn from staff and wire reports, "Investigators Comb French Crash Site Seeking Cause of Concorde Tragedy," CNN.com, July 26, 2000, https://goo.gl/xSrSVm.

Quotes by German Chancellor Gerhard Schröder are drawn from Roger Cohen, "The Concorde Crash: Grief at Home: Germans Led by Schröder in Mourning Their Dead," *New York Times*, July 27, 2000; see also Rebecca Allison, "Messages of Sympathy Pour in for Victims," *Guardian*, July 26, 2000.

The quote by Bishop Josef is drawn from Cohen, "The Concorde Crash," as is the reference to the *Bild* headline.

Information on the nationalities of the four non-German passengers is drawn from Orlebar, *The Concorde Story*, 7th ed., 145.

The quote by Jean-Claude Gayssot, French transport minister, is drawn from Allison, "Messages of Sympathy."

The quote by German actor Günter Pfitzmann and the anecdote about his last-minute change in travel plans is drawn from "Families Wiped Out in Crash," BBC News, July 31, 2000.

The quote by British Conservative leader William Hague is drawn from Allison, "Messages of Sympathy."

Quotes by Michael Goldfarb, former chief of staff of the Federal Aviation Administration, are drawn from "Crash of Air France Concorde Kills at Least 113," CNN, program aired July 25, 2000.

The quote by Air France Concorde pilot François Pradon is drawn from Jason Burke and Arnold Kemp, "BA Urged to Ground Concorde," *Observer*, July 30, 2000.

The quote by columnist Richard Ingrams is drawn from his "Beauty Is All in the Ear," *Observer*, July 30, 2000.

Information regarding British Airways decisions to cancel its Tuesday Concorde flights to and from New York and resume them on Wednesday is drawn from Orlebar, *The Concorde Story*, 7th ed., 145. The detail regarding the engines being inspected after the crash and the quote by British Airways Concorde pilot Mike Bannister are drawn from "BA Resumes Supersonic Flights," BBC News, July 26, 2000.

The quote by retired British Airways Concorde Captain John Hutchinson is drawn from Tony Snow, "BA Suspends Concorde Flights," *Independent*, July 24, 2000.

The substance of and details contained in English translations of the July 27, 2000, bulletin issued by the Bureau d'Enquêtes et d'Analyses pour la Sécurité de l'Aviation Civile are drawn from Orlebar, *The Concorde Story*, 7th ed., 145; see also "French Account of Concorde Crash," *New York Times*, July 28, 2000.

Background information and details on the role of the French judiciary, its tasks, and its authority regarding evidence gathered during the investigation of a crash are drawn from Orlebar, *The Concorde Story*, 7th ed., 148–49.

Quotes and observations by BEA investigator Yann Torres are drawn from the transcript of an interview with him conducted by Caroline Buckley, April 9, 2014, Aéroport de Paris–Le Bourget, for *Air Crash Investigation*.

The reference to the seven working groups charged with finding and analyzing evidence and information pertaining to the crash is drawn from ibid. and from "Organization of the Investigation," *BEA Report*, 15.

Quotes and observations by Jean-Louis Chatelain are drawn from the transcript of an interview with him conducted by Caroline Buckley, April 7, 2014, Nice, France, for *Air Crash Investigation*.

Information about the Dunlop Aircraft Tire Co. as the supplier of tires used on Concordes flown by British Airways is based on reporting in "Runway Debris Blamed for Concorde Crash," TireBusiness.com, August 17, 2000, https://goo.gl/c7ersc.

Descriptions of the crash site (both quoted and paraphrased), details about its size and location in relation to nearby roadways, the position of the aircraft upon impact and its heading prior to the crash, the dispersal of wreckage in the debris field, and the location of specific parts of the plane are all drawn from "The Accident Site: Description of Site and Plan," *BEA Report*, 67–69.

Details about the 1992 crash of Air Inter Flight 148 from Lyon to Strasbourg are drawn from Federal Aviation Administration, "Air Inter Flight ITF 148, Airbus A320-111, F-GGED," https://goo.gl/JZEkGv.

Quotes and observations by Alain Bouillard are drawn from the transcript of an interview with him conducted by Caroline Buckley, April 9, 2014, Aéroport de Paris–Le Bourget, France, for *Air Crash Investigation*.

Quotes and observations by Bob MacIntosh are drawn from the transcript of an interview with him conducted by Sonya Buyting, March 27, 2014, Hamilton, Ontario, for *Air Crash Investigation*.

The quote by an emergency worker at the scene of the crash is drawn from reports by Agence France-Presse included in Steven Erlanger, "The Concorde Crash: The Scene: A Symbol of France Crashes in Flames," *New York Times*, July 26, 2000.

Andras Kisgergely's account of taking photos of the flight with a friend, and quotes by him, are drawn from "Concorde Photos Sold for a Song," BBC News, July 27, 2000.

The account of public displays of mourning by Gonesse residents, as well as details about Michèle Fricheteau, is drawn from John Henley, "Concorde 'Downed by Burst Tyre,'" *Guardian*, July 29, 2000.

Excerpted quotes by Martin Woollacott are drawn from Martin Woollacott, "The Message of Concorde Is That We Can't Go Any Faster," *Guardian*, August 18, 2000.

4. From Dream to Reality

The paper-airplane anecdote is drawn from "Paper Airplanes Were Used to Test Concorde Wing Shapes," *Daily Mail*, September 27, 2007, as are quotes by former Concorde designer Alan Perry and by Peter Turvey, former senior curator at the Science Museum in Wroughton.

Historical details regarding the first rocket planes are drawn from Matthew A. Bentley, *Spaceplanes: From Airport to Spaceport* (New York: Springer, 2009), 7.

Details and background on the fifth Volta Conference are drawn from Stephan Wilkinson, "Mach 1 Assaulting the Barrier," *Air and Space Magazine*, December 1, 1990; see also John D. Anderson Jr., *A History of Aerodynamics: And Its Impact on Flying Machines* (Cambridge: Cambridge University Press, 1997), 401.

Descriptions of Ralph Virden's fatal test dive and the phenomenon of the Mach tuck are drawn from Wilkinson, "Mach 1 Assaulting the Barrier."

Quotes by author Christopher Orlebar about the structural stress on aircraft experiencing Mach tuck, so that the plane would "buffet and shake, sometimes to the point of structural failure"; details about test pilot Tony Martindale's experience of "strange effects when diving at such speeds"; and the quote about the tendency for Martindale's controls "to do the reverse of what was expected of them" are all drawn from Orlebar, *The Concorde Story*, 12–13.

Details concerning the death of test pilot Geoffrey de Havilland Jr. and his plane, as well as the excerpted quote about the speed reached by his plane prior to the crash, are drawn from "De Havilland Reached Speeds 'in Excess of World Record,'" *Observer*, September 28, 1946.

Details and observations of experimental test pilot George E. Cooper's in-flight experience are drawn from Wilkinson, "Mach 1 Assaulting the Barrier," as are details about Charles "Chuck" Yeager becoming the first pilot to surpass the speed of sound.

General background on and details about the formation of the Supersonic Transport Aircraft Committee, its goals, and its key personnel are drawn from Orlebar, *The Concorde Story: 21 Years in Service*, 22; see also Rhodri Owen, "Concorde Engineer Sir Morien Morgan Remembered on Centenary," *BBC Wales*, December 20, 2012.

The excerpted quote about Britain's political climate in 1962 and the push to build a supersonic aircraft for commercial travel is drawn from Orlebar, *The Concorde Story*, 7th ed., 41; insight into France's efforts, starting in the 1950s, to build a series of supersonic research aircraft is drawn from ibid., 31. Information about the French Super Caravelle model displayed at the Paris Air Show of 1961, its similarities to British BAC 223, and the corporate history of the Bristol Aeroplane Company is drawn from ibid., 32. Background on and information about which nations and companies would take on various aspects of Concorde's research, design, and development, as well as details of the plans calling for completion and testing of two prototypes and their projected development costs, are drawn from ibid., 35. Orlebar's quote about "totally unforeseen side effects" is drawn from ibid., 38. His observation about scrutiny of Concorde's design and safety features by civil aviation authorities in three countries is drawn from ibid., 40.

Brian Trubshaw's quotes regarding the successful development of Concorde and his reflections on the difficulty of the undertaking are drawn from his *Concorde: The Inside Story* (Thrupp: Sutton Publishing, 2000), Introduction, xiii; for details on national rivalries, see ibid., 22. The account of arguments between French and English pilots about matters large and small, including visual displays on the instrument panel and related quotes, are drawn from ibid., 54.

Details about the 250 British Airways engineers who helped to design Concorde's airframe and engines are drawn from British Airways, "Celebrating Concorde: About Concorde," https://goo.gl/HYxTN5.

The fact that the two small Concorde fleets accumulated more hours of supersonic flight during the life of the Concorde project than the sum total of all the world's military planes combined was reported by Richard Seebass, "Introduction: History and Economics of and Prospects for Supersonic Transport," *Fluid Dynamics Research on Supersonic Aircraft*, NATO Research and Technology Organization (November 1988): 1.

Biographical information about mathematician Johanna Weber, her education and career, and her collaboration with Dietrich Küchemann are drawn in part from Martin Childs, "Johanna Weber: Mathematician and Aerodynamics Expert, Whose Work on Wing Design Played a Key Role in Developing Concorde," *Independent*, November 30, 2014; see also John Green, "Obituary—Dr. Johanna Weber," *Royal Aeronautical Society News and Expertise,* January 12, 2015.

The quote by author and test pilot Eric Brown about Dietrich Küchemann's presentation on the slender wing concept is drawn from Eric Brown, *Wings on My Sleeve: The World's Greatest Test Pilot Tells His Story* (London: Weidenfeld & Nicolson, 2008), 177–78.

Information on Concorde prototypes used for testing and the facility used for that purpose at the RAE complex in Farnborough is drawn from Trubshaw, *Concorde*, 34.

Information about the 10 Concordes assembled at Filton and the 10 others at Toulouse is drawn from Aerospace Bristol, "Concorde Timeline: Design," https://goo.gl/A6A93N.

Brian Trubshaw's quote about the fate of unsold Concordes, as well as descriptions of the prototypes and preproduction planes, is drawn from Trubshaw, *Concorde*, 113–14. Information about specially constructed flight simulators and the related quote by the author is drawn from ibid., 55.

Excerpted descriptions of the March 2, 1969, flight of French Concorde prototype 001 are drawn from David Fairhall, "Concorde Takes to the Skies," *Guardian*, March 3, 1969.

The quote by Concorde pilot André Türcat is drawn from Orlebar, *The Concorde Story: 21 Years in Service*, 47.

Details on the successful test flight of Concorde prototype 002 are drawn from David Fairhall, "Concorde 002 Flies in the Severn Sun," *Guardian*, April 10, 1969.

The quote by Concorde pilot test Brian Trubshaw is drawn from Paul Lewis, "Brian Trubshaw, 77, Dies; Tested Concorde," *New York Times*, March 28, 2001.

Information about the Tupolev Tu-144, the crash at the Paris Air show in 1973, the ensuing controversy, and its eventual use as a cargo plane is drawn from Yuri Zarakhovich, "The Concordski," *Time*, August 7, 2000. Dates regarding the Tu-144's maiden flight and its first supersonic flight are drawn from Orlebar, *The Concorde Story*, 7th ed., 69.

Brian Trubshaw's quotes about sonic booms are drawn from his *Concorde*, 1. Material on anticipated sales of Concorde and potential overland routes for Concorde, as well as the anecdote about the Imelda Marcos shopping trip, are drawn from ibid., 1, 129, 128. Details on the terms of the sales of Concordes to British Airways and Air France in 1972 are drawn from ibid., 132–33; details on the start of Concorde's commercial service in Britain and France on January 21, 1976, and the congratulatory quote by Queen Elizabeth II are drawn from ibid., 110.

The commercial-use history of Concorde F-BTSC, including lease and purchase dates by Air France, is based on information found in "Concorde the Fleet, Aircraft 203: F-BTSC," https://goo.gl/CQUtM3.

Names of various celebrities who flew Concorde over the years and related details are drawn from "Concorde's Celebrity Passengers," *Daily Mail*, February 26, 2018.

Information about Concorde charter operators, special flights, and the August 1995 round-the-world flight is drawn from Trubshaw, *Concorde*, 144, 148. Details about the British Airways–owned Concorde that set a transatlantic record on February 7, 1996, and the related quote by BA flight manager Mike Bannister regarding the BA Concorde fleet, are drawn from ibid., 147.

5. A Trail of Evidence

Quotes and observations by Bob MacIntosh are drawn from the transcript of an interview with him conducted by Sonya Buyting, March 27, 2014, Hamilton, Ontario, for *Air Crash Investigation*.

Details about the aftermath of the crash, including the melting plastic fixtures on the neighboring hotel exterior and the amount of water and emulsifiers used to extinguish the blaze, are drawn from "Fire," *BEA Report*, 88.

Quotes and observations by Alain Bouillard are drawn from the transcript of an interview with him conducted by Caroline Buckley, April 9, 2014, Aéroport de Paris–Le Bourget, France, for *Air Crash Investigation*.

The location of the cockpit and the flight crew's bodies is based on information in "Survival Aspects," *BEA Report*, 88.

The specified length and weight of the plane's four engines is based on information in Gordon Roxburgh, "Concorde Technical Specs, Power Plant," *Concorde SST*, https://goo.gl/zQCZGV.

The estimated quantity of fuel (60 kilograms per second) released by the leak is drawn from "Estimation of Fuel Flow," *BEA Report*, 119. The widths and lengths of runways 26R and 26L, and the size and number of the concrete slabs that make up runway 26R, are drawn from "Aerodrome Information," *BEA Report*, 39–40.

The method employed to construct a grid of the runway using the concrete slabs is noted in "Wreckage Information," *BEA Report*, 59. Information about the location of the pieces of the plane's water deflector and pieces of tire tread is drawn from "Water Deflector" and "Pieces of Tire," *BEA Report*, 59–60.

The weight of the largest piece of tire debris found on the runway is drawn from "Pieces of Tire," *BEA Report*, 60. The size and location of the kerosene stain and soot deposits on the runway are drawn from "Soot Deposits on Runway," *BEA Report*, 64–65.

Details about the broken safety light, and the likely cause of the damage, are drawn from "Lighting," *BEA Report*, 62. Details about the scorched grass along the left side of the runway are drawn from "Soot Deposits on Runway," *BEA Report*, 65. Details about the tire tracks on the runway are drawn from "Tire Tracks," *BEA Report*, 62–63.

The size, condition, and location of the fuel-tank hull fragment are drawn from "Structural Elements," *BEA Report*, 61. Descriptions of debris found past the end of the runway and up to the crash site are drawn from "Between Runway 26 Right and the Accident Site," *BEA Report*, 66. The location and description of the strip of metal are drawn from "Piece of Metal," *BEA Report*, 61.

6. Too Late

Information in this chapter about the retrieval of the flight data recorder (FDR), cockpit voice recorder (CVR), and quick access recorder (QAR) from the crash site, including when and how the recorders were located, how they were placed under seal, where they were taken for analysis, the condition of their casings, the make of the recorders, and the condition of the tapes and casings, is drawn from "Flight Recorders," *BEA Report*, 41–43.

Unless otherwise indicated, the times cited for events described in this chapter are drawn from Appendix 2, "CVR Transcript" and "Analysis," *BEA Report*, 158–64.

Details about the BEA technician who donned an oxygen mask and goggles during the hunt for the recorders in the wreckage are drawn from an online transcript of the broadcast "What Brought Down the Concorde: *Dateline* Reexamines the Last Moments of the 2000 Flight That Proved Fatal," NBC News, February 1, 2010, https://goo.gl/A6n2rj.

Information about the tests, cleaning, and repair work performed on the recorders and the use of special software to produce improved readouts is drawn from "Flight Recorders," *BEA Report*, 40–43.

Quotes and observations by Alain Bouillard are drawn from the transcript of an interview with him conducted by Caroline Buckley, April 9, 2014, Aéroport de Paris–Le Bourget, France, for *Air Crash Investigation*.

Details about Air France pilots who knew Flight 4590's cockpit crew verifying that the voices heard on the CVR are those of Christian Marty, Jean Marcot, and Gilles Jardinaud are drawn from information contained in "Cockpit Voice Recorder," *BEA Report*, 41. Information about the types of background noises recorded on the CVR is drawn from "The Flight Up Until the Loss of Thrust on Engine 1," *BEA Report*, 163.

Problems encountered due to the frequency of the sampling of engine data on the FDR, and methods used to transfer the data to a new memory card, are described in "Flight Recorders," *BEA Report*, 172.

Quotes and observations by BEA investigator Yann Torres are drawn from the transcript of an interview with him conducted by Caroline Buckley, April 9, 2014, Aéroport de Paris–Le Bourget, France, for *Air Crash Investigation*.

Details on the April 4, 1977, crash of Southern Airways Flight 242 are drawn from the National Transportation Safety Board's official report on the accident, *NTSB-AAR-78-3, Aircraft Accident Report—Southern Airways Inc., DC-9-31, N1335U, New Hope Georgia, April 4, 1977* (Washington, DC: National Transportation Safety Board Bureau of Accident Investigation, 1978).

Details regarding engine failure and the forced ditching of US Airways Flight 1549 on January 15, 2009, are drawn from National Transportation Safety Board, "History of the Flight" and "Injuries to Persons," *Loss of Thrust in Both Engines After Encountering a Flock of Birds and Subsequent Ditching on the Hudson River, US Airways Flight*

1549, Airbus A320-D214, N106US, Weehawken, New Jersey, January 15, 2009, Accident Report NTSB/AAR-10/03 (Washington, DC: National Transportation Safety Board, May 4, 2010), 1–5, 6.

Information about the loss of power in the number 2 engine is drawn from "The Flight Up Until the Loss of Thrust on Engine 1," *BEA Report*, 162. Information about when the fire alarm sounded and the flight engineer's announcement that he was shutting down engine number 2 is drawn from ibid., 163. Information regarding the time when engines number 2 and 1 suffered their first loss of thrust is drawn from ibid., 161. Details of critical events that occurred on the runway after Concorde's tire burst, including the drop in power, the leftward yaw, Captain Marty's decision to rotate the plane, and the speed at which the plane was traveling at that time, are drawn from ibid., 161–62. The plane's reduction of thrust by 50 percent, and the detail that most of the power was being generated by engines number 3 and 4, is drawn from ibid., 162.

Information about reconstruction of the delta wing using the remains of the fuel tanks and engines is drawn from "Reconstruction of the Wing and Examination of the Debris," *BEA Report*, 81–82.

Investigators' conclusion that fire damage to Concorde's wing and flight controls would have caused the plane to crash even if all four engines had been operating normally is drawn from "Loss of Control of the Aircraft," *BEA Report*, 165.

7. A History of Blowouts

Quotes and observations by Bob MacIntosh in this chapter are drawn from the transcript of an interview with him conducted by Sonya Buyting, March 27, 2014, Hamilton, Ontario, for *Air Crash Investigation*, as well as from email correspondence between the author and MacIntosh, December 4, 5, 7, and 8, 2017.

Quotes and observations by Alain Bouillard in this chapter are drawn from the transcript of an interview with him conducted by Caroline Buckley, April 9, 2014, Aéroport de Paris–Le Bourget, France, for *Air Crash Investigation*.

Quotes by Joerg Horny and details on lawsuits filed by relatives of crash victims are drawn from Andrew Alderson, David Harrison, and Julian Coman, "Concorde Crash Families Sue for £3 Million Each," *Telegraph*, July 30, 2000.

Details regarding Flight 54's tire burst mishap on June 14, 1979, are drawn from "Events Which Caused Structural Damage to Tanks" and "Appendix 5: Previous Events," *BEA Report*, 94 and 1 of Appendix 5. See also Thomas Grubisich and Stephanie Mansfield, "Concorde Scare," *Washington Post*, June 15, 1979, the source of the quote by Patrick T. Chitwood, an airport operations officer who witnessed the Flight 54 incident.

The summary of the findings of the BEA investigation undertaken after the Flight 54 incident is drawn from "Event on 14 June 1979 at Washington," *BEA Report*, 96.

Details on and quotes from the letter that the NTSB sent to French air safety officials in November 1981 are drawn from Richard Witkin, "FAA Troubled by Concorde Tire Blowouts," *New York Times*, November 15, 1981.

The reference to Air France's discontinuation of the use of retread tires starting in January 1996 is drawn from "Figure 45: History of Concorde Tire Events," *BEA Report*, 94.

The references to 57 documented tire mishaps and their causes since Concorde had begun commercial flights in 1976, including deflated or burst tires and loss of tread, are based on information in "Nature of Events," *BEA Report*, 93; see also "Figure 45: History of Concorde Tire Events," *BEA Report*, 94.

The reference to Concorde tire bursts occurring at a rate of one for every 4,000 flying hours, or about 60 times more frequently than the rate for other long-haul aircraft such as the Airbus A340, is drawn from "Points Related to Tires and Structural Damage," *BEA Report*, 146, and also "Concorde Tires 'Had Burst 57 Times Before,'" *Daily Mail*, July 26, 2001.

Information on how frequently Concorde tires were changed is drawn from David Ruppe, "Concorde Crash Raises Safety Questions: Experts," ABC News, July 30, 2000.

Information on and analysis of the specific type and number of tire events by year over the 17-year period from 1982 to 1999 is drawn from "Figure 45: History of Concorde Tire Events," *BEA Report*, 94.

Information on the October 25, 1993, tire burst incident involving a British Airways Concorde taking off from London's Heathrow Airport is drawn from "Events Which Caused Structural Damage to Tanks," *BEA Report*, 95. Examples of types of secondary damage resulting from burst tires are drawn from ibid., 94.

Information about the number of daily runway inspections at Charles de Gaulle Airport and the time of day the inspections were carried out is drawn from "Runway Inspections," *BEA Report*, 40. See also "Runway Was Not Checked for Doomed Concorde," *Guardian*, September 1, 2000.

Times of day that inspections were carried out on the day of the accident are drawn from "The Inspections on 25 July 2000," *BEA Report*, 41. The fact that runway 26R had not been fully inspected for more than 12 hours prior to Concorde's takeoff at 4:42 p.m. is drawn from ibid.

Information regarding the two planes that had taken off prior to Flight 4590—a Continental Airlines DC-10 and an Air France Boeing 747—is drawn from "Metallic Strip Found on the Runway," *BEA Report*, 103. The August 30, 2000, date given for the Continental DC-10's anticipated return to Charles de Gaulle is drawn from ibid.

Background on and details of the BEA investigator's two-week-long hunt for the plane that might have dropped the metal strip are based on information provided by email correspondence between the author and Bob MacIntosh, December 2, 2017.

General information on and specific details of the tests carried out at the Goodyear technical center, as well as their results, are drawn from "Experimental Tests," *BEA Report*, 97–98. Descriptions of the tire tests conducted at the Aeronautical Test Center in Toulouse, the BEA technicians' tests of tire chunks at the Rubber and Plastics Research and Test Laboratory, and the results of those tests are drawn from "Tests Carried Out at the CEAT," *BEA Report*, 101. The results of the analysis of the primer paint on the DC-10's engine and other samples collected by BEA investigators are drawn from "Examination of Samples Taken from N 13067," *BEA Report*, 107.

Information about the check of Concorde's maintenance records and the discovery that a spacer in the left main landing gear had not been reinstalled during an Air France repair that took place between July 17 and 21, 2000, is drawn from "Absence of the Spacer on the Left Main Landing Gear," *BEA Report*, 148–49. Details of the decision to replace Concorde's whole left-main landing gear bogie beam with a new beam is drawn from "Maintenance Operations," *BEA Report*, 149. The fact that a prescribed tool needed to replace the bogie beam was not used by Air France mechanics is drawn from "Work Performed," *BEA Report*, 150.

The reference to the dates of four flights (July 21, 22, 23, and 24) undertaken by Concorde F-BTSC after the left-main landing gear bogie had been replaced, minus the spacer, is based on information in "Maintenance," *BEA Report*, 21.

The reference to the FDR data check that revealed that brake temperatures were normal for both the left and right bogies while the plane was taxiing to its holding position on runway 26R is drawn from "Study of the Beginning of the Flight," *BEA Report*, 155. References to FDR data showing that Captain Marty did not need to use the right rudder to stay on course before the tire burst and that no tire skid marks were found on the runway before that point are drawn from ibid.

The observation that BEA investigators found considerable fault with maintenance work performed on Concorde F-BTSC, calling the omission of the spacer a "grave error," is drawn from "Functioning of Maintenance," *BEA Report*, 170.

Statements regarding BEA research on the bogie's pulling to the left are based on information found in "Possible Consequences on the Landing Gear of the Absence of the Spacer," *BEA Report*, 152–53.

The reference to the withdrawal of the airworthiness certificate for both Concorde fleets and the last-minute arrangements made for passengers when British Airways cancelled a flight from London to New York as the plane stood on the runway at Heathrow Airport is drawn from Rachel Donnelly, "BA Concorde Fleet Grounded Following Notice of Withdrawal of Airworthiness Certificate," *Irish Times*, August 16, 2000.

Information regarding the legal claim filed against Continental Airlines by Air France and its insurer, as well as the quote by a spokeswoman for Air France, are drawn from Severin Carrell, "Air France Sues US Airline over Concorde Crash," *Independent*, September 27, 2000.

Information on when and where Concorde F-BTSC ran over the metal strip and the plane's estimated speed at that time is based on related information in "The Flight Up Until the Loss of Thrust on Engine 1," *BEA Report*, 161.

8. An Invisible Shock Wave

Quotes and observations by Bob MacIntosh in this chapter are drawn from the transcript of an interview with him conducted by Sonya Buyting, March 27, 2014, Hamilton, Ontario, for *Air Crash Investigation*, as well as from email correspondence between the author and MacIntosh, December 4, 5, 7, and 8, 2017.

Quotes and observations by Alain Bouillard in this chapter are drawn from the transcript of an interview with him conducted by Caroline Buckley, April 9, 2014, Aéroport de Paris–Le Bourget, France, for *Air Crash Investigation*.

Quotes and observations by BEA investigator Yann Torres are drawn from the transcript of an interview with him conducted by Caroline Buckley, April 9, 2014, Aéroport de Paris–Le Bourget, for *Air Crash Investigation*.

The size and description of the large, square piece of the number 5 fuel tank found on the runway are drawn from "Examination of the Pieces of the Tank," *BEA Report*, 109.

The weight of the tire fragment found on the runway is drawn from "Pieces of Tire," *BEA Report*, 60; the estimated speed of the tire fragment is drawn from the *Air Crash Investigation* transcript of the interview with Alain Bouillard.

The reference to the effects of a shock wave phenomenon occurring in fuel tanks in military aircraft under fire is drawn from Christopher Orlebar, *The Concorde Story*, 7th ed., 163. The explanation of the effects of acceleration on the fuel in the tanks of Concorde F-BTSC on the day of the crash is drawn from ibid. The quote beginning "If the undercarriage could have been retracted" is drawn from ibid., 166. Details about ignition tests conducted by the BEA at Warton can be found in ibid., 165. Quotes by the author regarding ignition test results are drawn from ibid., 165–66.

The reference to the unusual nature of the shock wave phenomenon in the tanks of civilian aircraft is drawn in part from "Destruction of the Lower Part of Tank Five," *BEA Report*, 167.

Details pertaining to and descriptions of tests undertaken by the BEA investigators are based in part on information found in "Tank Rupture Mechanism," *BEA Report*, 112–17.

The quote by Jonathan Glancey about the electrical arc from the brake cooling fan as the likely source of the spark that lit the Concorde fire is drawn from Jonathan Glancey, *Concorde: The Rise and Fall of the Supersonic Airliner* (London: Atlantic Books, 2015), 194.

The quote from the official FAA account of the ignition source of the leaking fuel is drawn from "Lessons Learned from Civil Aviation Accidents, Additional Post Accident Testing," *Federal Aviation Administration*, https://goo.gl/uTwySS.

9. Delays and Headaches

The four quotations from Antoine de Saint-Exupéry can be found, in the order given, in his *Wind, Sand, and Stars*, trans. Louis Galantiére (New York: Harcourt Brace and Company, 1939), 161, 73, 176, 37. Biographical details regarding Saint-Exupéry are drawn from Isabelle de Courtivron, "At Home in the Air," *New York Times*, January 8, 1995.

Quotes and observations by Jean-Louis Chatelain in this chapter are drawn from the transcript of an interview with him conducted by Caroline Buckley, April 7, 2014, Nice, France, for *Air Crash Investigation*.

Details on the flight crew's ratings and professional background and qualifications can be found in "Flight Crew," *BEA Report*, 18–20.

The description of the instrument panels in Concorde cockpits is based in part on information provided by Orlebar, *The Concorde Story: 21 Years in Service*, 85, and from photographs from ibid., 88, 46–47, 127, 134.

Information about the duties of the Air France flight dispatcher and support staff on duty the day of the crash, the use of a computer program to assist in certain aspects of flight preparation, and the time that work was carried out, as well as manual calculations used to forecast takeoff weight and the fuel required for the flight, is based on details and statements provided in "Flight Planning," *BEA Report*, 88.

References to the Flight 4590 crew's activities prior to departure, the timing of that activity, their arrival at the flight departure center to collect and study the flight dossier, and their communication with the dispatcher are based on information provided in "Flight Departure," *BEA Report*, 88.

Information regarding Captain Marty's routine prior to takeoff on July 25, 2000, including his responsibility for studying the flight dossier and signing the fuel load sheet, is based on statements in "Flight Departure," *BEA Report*, 88–89.

Information about the dispatcher's preflight duties and activities on the day of the crash, including drawing up a load plan and a forecast for the final weight of baggage, calculating the CG forecast, and handing that information to the flight crew to be reviewed by the captain, is based on related information provided in "Traffic," *BEA Report*, 89.

References to and descriptions of the dispatcher's decision to impose a performance penalty and reduce by 2.5 percent the maximum weight for Concorde F-BTSC due to the reported problem with the plane's engine 2 thrust reverser are based on information provided in "Flight Planning," *BEA Report*, 89–90. References to the dispatcher's exchange with the duty officer regarding the performance penalty, the effect on the plane's estimated weight allowance, and possible remedial strategies proposed by both the dispatcher and the duty officer are based on information provided in "Flight Planning," *BEA Report*, 89–90.

References to the communication between the flight crew and the dispatcher prior to the flight's departure, the flight crew's decision to take over flight preparation, and Captain Marty's order to repair the thrust reverser are drawn from information provided in "Flight Planning," *BEA Report*, 89–90.

Details about the overnight change in status for Concorde F-BTSC from reserve to service as Flight 4590 can be found in "Aircraft Information: Maintenance," *BEA Report*, 22.

The reference to the total number of bags loaded, 122 (versus the 103 that were listed on the load sheet given to the crew), as well as the average estimated weight of 20.7 kilogram each, reflects information provided in "Weight," *BEA Report*, 32.

References to surface winds reported in the area of runway 26R are based on information in "Meteorological Conditions: Situation at the Aerodrome," *BEA Report*, 36.

The three V-speeds selected by Captain Marty prior to takeoff, "V1:150 kt, VR: 198 kt, and V2: 220 kt," are listed in "History of the Flight," *BEA Report*, 17.

Details regarding estimated weights for persons on board—88 kilograms (194 lb) for each man, 70 kilograms (154 lb) for every woman, and 35 kilograms (77 lb) for each child—and also for each piece of luggage—20.7 kilograms (45.6 lb)—are drawn from "Weight," *BEA Report*, 32.

Details about the engine repair performed on Concorde F-BTSC as it sat parked at the Air France gate, and its completion at "14 h 16 min and 11 secs," are drawn from "Maintenance" and "Analysis," *BEA Report*, 22, 158; see also Anne Swardson and Don Phillips, "Crash Probe Centers on Jet Engine," *Washington Post*, July 27, 2000.

Details about low-pressure systems in Europe, local temperatures at the airport, and visibility on the day of the crash can be found in "Meteorological Conditions: Situation at the Aerodrome," *BEA Report*, 36.

Information about the plane's assigned V speeds is drawn from "History of the Flight," *BEA Report*, 17.

Information regarding the various factors affecting the maximum performance on Concorde as stated in the Concorde "operating manual" is drawn from "Take-off Performance," *BEA Report*, 33.

The explanation regarding the impossibility of setting actual weights due to the use of fixed average weights is drawn from footnote 8, *BEA Report*, 31.

Concorde's recommended maximum structural weights, including taxi weight (186,880 kg) and takeoff weight (185,070 kg), are found in "Takeoff Performance," *BEA Report*, 33.

Information regarding the weight of the fuel in the plane's tanks upon fueling—94,800 kilograms, or 94.8 metric tons (104 standard tons)—and the time Concorde fuel tanks were filled ("13 h 55") are drawn from "Fuel," *BEA Report*, 27.

Concorde's recommended maximum structural weights (specified in its operating manual), including taxi weight (186,880 kg) and takeoff weight (185,070 kg), are found in "Takeoff Performance," *BEA Report*, 33.

The fact that the crew took off with the plane's center of gravity at 54 percent can be found in "The Fuel in Tank 5," *BEA Report*, 117: "Before line-up, the crew carried out fuel transfer so as to bring the CG up to 54% for takeoff."

The statement regarding Concorde F-BTSC's structural weights being within operational limits prior to takeoff is drawn from "Findings," *BEA Report*, 174.

Information regarding the estimated weights assigned to baggage and persons on board and the final baggage count is drawn from "Weight," *BEA Report*, 32.

The estimated total weights of the plane prior to takeoff (both original and revised) and the plane's center of gravity are drawn from factual information provided in "History of the Flight," "Weight" and "CG," *BEA Report*, 17, 31, 32.

The discussion of the location of the "usual" or prescribed center of gravity position for all Concordes is based in part on analysis provided in Orlebar, *The Concorde Story*, 7th ed., 148.

The time of the tow truck arrival and the westward-facing position of the plane at the time are drawn from Appendix 2, "CVR Transcript," *BEA Report*.

The quoted description of passengers on Concorde charter flights is drawn from Sally Armstrong, *Vintage Champagne on the Edge of Space: The Supersonic World of a Concorde Stewardess* (London: History Press, 2015), 73. The quoted excerpt regarding the British Airways Concorde captain deciding against taking an en route diversion to refuel can be found in ibid., 56.

Descriptions of meals served on Concorde are drawn from Patrick Skene Catling, "An Elegy for Concorde, the Most Beautiful Airliner of All Time," *Spectator*, November 14, 2015.

The ages and job descriptions of the six crew members working in Flight 4590's passenger cabin are drawn from "Personnel Information," *BEA Report*, 18–20.

Quotes by Stéphane Garcia about his late brother, flight attendant Hervé Garcia, are drawn from Pierre Verdy, "Concorde: The Trauma Still Alive 12 Years Later for Civil Parties," *Agence France-Presse*, December 6, 2012.

Biographical and professional details about cabin crew member Brigitte Kruse are drawn from "Flight Crew," *BEA Report*, 18–20; Derek Scalley, "Chancellor to Attend Cologne Service," *Irish Times*, July 29, 2000.

Biographical and professional details about cabin crew member Virginie-Huguette Le Gouadec are drawn from "Flight Crew," *BEA Report*, 20–21; and from Laurent Benayou, "Virginie Le Gouadec, the Chef de Cabin: She Was Flying in Castres and Revel," *La Depeche*, July 27, 2000.

Biographical and professional details about cabin crew member Patrick Chevalier are drawn from "Flight Crew," *BEA Report*, 20–21; and from "A Concorde Steward Lived in Bouillancy," *Le Parisien*, July 28, 2000.

Biographical and professional details about cabin crew member Anne Porcheron are drawn from "Flight Crew," *BEA Report*, 20–21; and from Véronique Escolano, "At the SOS Village, Brothers and Sisters Together Find a Family," *Ouest France Student News*, May 17, 2015.

Biographical and professional details about Florence Eyquem-Fournel are drawn from "Flight Crew," *BEA Report*, 18–20; and from Xavier Massé, *Avion Concorde* (Paris: Nouvelles Éditions Latines, 2004), 104.

Biographical and professional details about cabin crew member Hervé Garcia are drawn from "Flight Crew," *BEA Report*, 18–20; and from Verdy, "Concorde: The Trauma Still Alive."

Details about the training of Air France and British Airways Concorde cabin crews is based on information in "Concorde Back in the Skies: Cabin Crews," Air France Press Office, October 2001, https://goo.gl/GmcsAE; and Armstrong, *Vintage Champagne on the Edge of Space*, 35–38. The quote by Sally Armstrong about the "coolness of character" required of Concorde flight attendants is drawn from ibid., 31.

10. Three Seconds of Mayhem

Quotes concerning Christian Marty by Bernard Pedamon and Pierre-Jean Loisel are drawn from John Henley, "Skilled Pilot Who 'Did His Best,'" *Guardian*, July 27, 2000.

Insights into the character of Jean Marcot, and his popularity among Air France coworkers and related details, are drawn from Nelly Terrier, "A Crew Who Lived Their Passion," *La Parisienne*, July 27, 2000.

Details about Christian Marty's fondness for hiking and biking during stopovers are drawn from Thomas Sancton, "The Pilot," *Time*, July 30, 2000.

Quoted comments about Christian Marty by Claude Bouvier-Muller are drawn from Sancton, "The Pilot," and Thomas Sancton, "A Man at Home with Danger," *Time*, August 7, 2000.

Christian Marty's quote about his determination to rely on his own strength to achieve his goal of windsurfing across the Atlantic is drawn from Nadel, "A Tribute to Christian Marty."

Christian Marty's quote about the notion of an honorable defeat is drawn from Sancton, "A Man at Home with Danger."

Unless otherwise stated, all quotes in this chapter by pilots Christian Marty, Jean Marcot, and Flight Engineer Gilles Jardinaud and air traffic controller Gilles Logelin, as well as quotes by unidentified flight attendants and the dispatcher, are drawn from Appendix 2, "CVR Transcript," *BEA Report*, n.p., as are the times cited for specific events leading up to and during the in-flight emergency prior to the crash.

Quotes and observations by Jean-Louis Chatelain in this chapter are drawn from the transcript of an interview with him conducted by Caroline Buckley, April 7, 2014, Nice, France, for *Air Crash Investigation*.

The interpretation of the effects of air movement of 3 knots is based on "Beaufort Wind Scale," Storm Prediction Center, NOAA/National Weather Service, https://goo.gl/WhzczP.

The quotes by Sally Armstrong in which she compares Concorde to a "caged bird" and describes the excitement of takeoff are drawn from her *Vintage Champagne on the Edge of Space*, 57.

The 38-second duration of normal conditions during the initial takeoff run of Flight 4590 is drawn from "Crew Actions," *BEA Report*, 165: "During the first thirty-eight seconds of the takeoff, the crew were in a perfectly normal situation."

The specified 4.5-kilogram (9.9-lb) weight of the piece of tire that struck the number 5 fuel tank is drawn from "Findings," *BEA Report*, 174.

The quote describing Concorde as "sublime and magisterial" is from Glancey, *Concorde*, 4.

The explanation of the combined effects of the tire blowout, fuel release. and resulting fire, which affected the plane's heading, is based on information provided in "The Flight Up Until the Loss of Thrust on Engine 1," *BEA Report*, 161: "The heading change was probably the result of a combination of the tire burst and the aerodynamic disturbance due to the fuel leak and fire."

The timing of Captain Marty's use of rudder deflection preceding the engine surging event as recorded by engine parameters can be found in ibid.

The reported time that the engine "go lights" blinked on and off (14:43:16 to 14:43:18) can be found in "Data Readout," *BEA Report,* 130. The timing of the go lights blinking off as coinciding with First Officer Marcot's shouted warning, "Watch out," can be found in "The Flight Up Until the Loss of Thrust on Engine 1," ibid., 161.

The operational status of the number 2 engine at shutdown can be found in "Data Readout," and "Engine 2," *BEA Report,* 131, 133.

The change in the thrust lever setting to idle by Captain Marty can be found in "Engine 2," *BEA Report,* 133.

The quote from retired Concorde captain John Cook is drawn from Paul Kelso, "The First You Know of a Fire Is a Bell and a Red Light," *Guardian*, July 27, 2000.

The reference to the reduction of engine power to 50 percent and the source of thrust at that time coming primarily from engines number 3 and 4 can be found in "The Flight Up Until the Loss of Thrust on Engine 1," *BEA Report,* 162.

The speed of Concorde F-BTSC at rotation (183 knots) is drawn from "Findings," *BEA Report,* 175.

The description of the landing gear lever in the Concorde cockpit can be found in "Landing Gear Retraction," *BEA Report,* 22.

Speculations by Cathy Cooper on how the Concorde flight attendants might have responded to the emergency on Flight 4590 are based on email correspondence between Cooper and the author, December 17 and 19, 2017.

Details about Southern Airways Flight 242 are drawn from the National Transportation Safety Board's official report on the accident, *NTSB-AAR-78-3, Aircraft Accident Report—Southern Airways Inc., DC-9-31, N1335U, New Hope, Georgia, April 4, 1977*; see also Samme Chittum, *Southern Storm: The Tragedy of Flight 242* (Washington, DC: Smithsonian Books, 2018).

Information regarding the toilet smoke detection alarm heard at 4:43:32 p.m. and the infiltration of smoke into the air conditioning system, as well as the timing of the fire and the smoke detection alarms, is drawn from "Toilet Smoke Alarm," 138, and Appendix 3, "Alarms," *BEA Report,* 1.

Quotes and observations by Bob MacIntosh in this chapter are drawn from the transcript of an interview with him conducted by Sonya Buyting, March 27, 2014, Hamilton, Ontario, for *Air Crash Investigation*.

Quotes and observations by Alain Bouillard in this chapter are drawn from the transcript of an interview with him conducted by Caroline Buckley, April 9, 2014, Aéroport de Paris–Le Bourget, France, for *Air Crash Investigation*.

The quote from an anonymous witness describing the plane falling "like a dead leaf" is drawn from Swardson and Phillips, "Crash Probe."

11. Closing the Case

Quotes and observations by Jean-Louis Chatelain in this chapter are drawn from the transcript of an interview with him conducted by Caroline Buckley, April 7, 2014, Nice, France; quotes and observations by Bob MacIntosh are drawn from the transcript of an interview with him conducted by Sonya Buyting, March 27, 2014, Hamilton, Ontario; quotes and observations by Alain Bouillard are drawn from the transcript of an interview with him conducted by Caroline Buckley, April 9, 2014, Aéroport de Paris–Le Bourget, France; all for *Air Crash Investigation*.

The three probable causes of the crash are listed in "Conclusion," *BEA Report*, 176; the reference to the flight crew's inability to identify the fire's origin is drawn from ibid., 181.

Air France president Jean-Cyril Spinetta's quote about the cause of the crash likely being an engine fire is drawn from Joseph Harriss, "The Concorde Redemption," *Air and Space Magazine* (September 2001): 1.

Quotes and references describing Concorde as being slightly overloaded but within operational limits are drawn from "Weight and Balance," *BEA Report*, 31, and "Conclusion," *BEA Report*, 174.

The quote stating that the flight would have had a negative outcome "even if all four engines had been operating" is drawn from "Loss of Control of the Aircraft," *BEA Report*, 165: "In any event, even if all four engines had been operating, the serious damage caused by the intensity of the fire to the structure of the wing and to some of the flight controls would have led to the rapid loss of the aircraft."

The official assessment of the decision made by the flight engineer to shut down engine number 2 is drawn from ibid., 166.

Information regarding the source of the metal strip is drawn from "Conclusion," *BEA Report*, 174.

The quote by journalist David Rose regarding the failure of the Air France ground crew to replace a spacer is drawn from David Rose, "Doomed: The Real Story of Flight 4590," *Observer*, May 12, 2001.

The quote by Paul-Louis Arslanian is taken from Harriss, "The Concorde Redemption," 2.

Christopher Orlebar's quote regarding the use of the left rudder during takeoff prior to the tire burst is drawn from Orlebar, *The Concorde Story*, 7th ed., 167.

Excerpts of commentary from *Le Figaro* and *Times of London* are drawn from Harriss, "The Concorde Redemption," 1.

Christopher Orlebar's quote regarding historical precedents that boded well for Concorde's return to service is drawn from Orlebar, *The Concorde Story*, 7th ed., 162, as is the reference to the £14 million cost of cabin refurbishment.

The reference to blue leather seats is found in Glancey, *Concorde*, 206.

The names of the Concorde insiders attending the November 2000 meeting, the location of the meeting at Gatwick Airport, and details about the flip chart are drawn from Orlebar, *The Concorde Story*, 7th ed., 162.

The description of the plan of action devised by participants in the November 2000 planning conference and the related quote by author Jonathan Glancey are drawn from Glancey, *Concorde*, 205.

The quote by Gérard Le Houx, an official with France's Direction Générale de l'Aviation Civile, is drawn from Harriss, "The Concorde Redemption," 2; the quote by Howard Berry and the detail about 100 engineers working on refitting Concorde are drawn from ibid., 2.

The number of fuel tank liners required to retrofit each aircraft is drawn from Glancey, *Concorde*, 206, as is the description of work required to install liners in Concorde fuel tanks.

The detail about airlines allowing customers to tour hangars during refitting is drawn from Philip Shenon, "So High, So Fast, So Fashionable," *New York Times*, November 17, 2002.

Quotes regarding the engineering involved in aircraft tire development are drawn from Joe Escobar, "Understanding the Basics of Aircraft Tire Construction and Maintenance," *Aviation Pros*, May 1, 2001.

Christopher Orlebar's explanation of the new NZG Michelin radial tire's qualities is drawn from Orlebar, *The Concorde Story*, 7th ed., 168.

Information about the NZG tire's special properties, including the reference to pieces of a blown tire not exceeding 1 percent of the tire's total mass, is drawn from "Concorde Lands Safely in New York on Revolutionary Michelin Tire," press release, Michelin North America, November 7, 2001, reproduced online by Defense Aerospace, https://goo.gl/tFStsP.

Details regarding NZG tire tests conducted at two locations, the results of those tests, and test methods are drawn from Orlebar, *The Concorde Story*, 7th ed., 168–69. See also "New Tires Give Hope for Concorde," BBC News, June 7, 2001.

Details regarding the May 2001 test flight of an Air France Concorde at a flight test center in Istres, France, are drawn from Orlebar, *The Concorde Story*, 7th ed., 169.

The quote by Pierre Desmarets, the director general of Michelin's aviation tires, is drawn from Peter Humi, "Michelin Shows 'Safer' Concorde Tire," CNN.com, July 5, 2001, https://goo.gl/uoAMh3.

The quote by Concorde commander Mike Bannister is drawn from Orlebar, *The Concorde Story*, 7th ed., 171.

The quote by Air France chairman Jean-Cyril Spinetta dedicating the flight to the 113 people who died in the crash is drawn from Suzanne Daley, "The Sleek Concorde Is Once More Carrying Passengers," *New York Times*, November 8, 2001. Details regarding Concorde's resumption of service on November 7, 2001, including Mayor Rudolph Giuliani's comments regarding flights to New York City and the quote by Sting concerning fearful travelers regaining their confidence, are drawn from ibid.

Sting's quote about his enthusiasm for Concorde and supersonic travel, as well as details about Concorde's departure from Heathrow Airport and the description of its takeoff, is drawn from Staff and Agencies, "Concorde Returns to Service," *Guardian*, November 7, 2001.

12. Final Farewells

The quote by Concorde pilot Jock Lowe is drawn from his "Veteran BA Pilot and Legend Jock Lowe Speaks Out," *Braniff Pages*, November 6, 2003, https://goo.gl/5URKsX.

The quote by Air France flight attendant Joelle Cornet-Templet is drawn from Charles Brenner, "Concorde's Demise Brings the French out in Flights of Fancy," *Times*, May 29, 2003.

The quote by Lord Sterling is drawn from Francie Grace, "Final Flight of the Concorde," *Associated Press and CBS News*, October 24, 2003, https://goo.gl/skCQ8M.

Quotes and observations by Jean-Louis Chatelain are drawn from the transcript of an interview with him conducted by Caroline Buckley, April 7, 2014, Nice, France, for *Air Crash Investigation*.

Quotes and observations by Bob MacIntosh are drawn from the transcript of an interview with him conducted by Sonya Buyting, March 27, 2014, Hamilton, Ontario, for *Air Crash Investigation*.

The description of Concorde by photographer Wolfgang Tillmans is drawn from his *Concorde* (Cologne: Walther König, 2017).

Details of the number and dates of incidents involving rudder delamination on Concorde, as well as what was required to fix the problem, are drawn from

Orlebar, *The Concorde Story*, 7th ed., 178–79. Details about an engine surging incident in 2003 on a Concorde transporting British Chancellor of the Exchequer Gordon Brown are drawn from ibid., 178. The observation that engine surges during supersonic flight were "rare" is drawn from ibid., 179. Quotes referring to the threat posed to Airbus's reputation by further Concorde incidents and how such incidents could constitute "another nail in Concorde's coffin" are both drawn from ibid., 179. The quote referring to the series of mechanical events that "at best looked untidy and at worst heralded future failures" is drawn from ibid., 178. The quotes regarding Airbus, "with its reputation on the line," is drawn from ibid., 186. The quote about Concorde pilot Les Brodie's confidence in Concorde's technical soundness is drawn from ibid., 187. Details regarding the proposal to establish a Heritage Flight program and Airbus's rejection of the plan are drawn from ibid., 188. The Orlebar quote regarding preparations for Concorde's farewell flights out of Heathrow Airport is drawn from ibid., 189.

Details about the last commercial flight of an Air France Concorde are drawn from Laurent Lemel, "Last Air France Concorde Leaves Paris for the Last Time," Associated Press, May 30, 2003, https://goo.gl/iGKRsD, as is the quote by Air France employee Jean-Pierre Lefebvre.

Quotes by an anonymous charter passenger and by Air France chairman Jean-Cyril Spinetta are drawn from Gordon Roxburgh, "Concorde Retires, Air France Retirement," *Concorde SST*, May 30, 2003, https://goo.gl/J3vCEE, as is the quote by Sebastian Weder, Concorde team commander.

The figure of £40 million as the cost of the proposed overhaul of British Airways' Concorde fleet is drawn from Edward Wong, "For Concorde, Economics Trumped Technology," *New York Times*, October 24, 2003.

Details about Donald L. Pevsner's involvement in Concorde charter flights, his friendship with Concorde pilot Jean Marcot, and his part in the two record-setting flights of 1992 and 1995 are drawn from John Milgrim, "Crash a Personal Loss for Friend of the Co-Pilot," *Times Herald Record*, July 26, 2000, updated December 14, 2010.

Quotes and opinions by Donald L. Pevsner are drawn from his "Betrayal of Concorde," Concorde Spirit Tours, April 19, 2017, https://goo.gl/ijSGxi.

The quote by author Jonathan Glancey criticizing the corporate mindset behind the decision to retire Concorde is drawn from his *Concorde*, 223–24.

The quote by Concorde buff John Cowburn is drawn from "End of an Era for Concorde," BBC News, October 24, 2003.

Details about the three Concorde flights arriving at Heathrow on October 23, 2003, and the celebratory sendoff for Concorde at JFK are drawn from ibid. and from Francie Grace, "Final Flight of the Concorde," Associated Press and CBS News, October 24, 2003; the latter is also the source of the quote by model Christie Brinkley.

The quote by actress Joan Collins is from "End of an Era for Concorde."

Quotes and observations by Concorde captain Michael Bannister and related details are drawn from Grace, "Final Flight of the Concorde."

Quotes by reporter Clyde Haberman are drawn from his "NYC: An End at Last to the Battle of Concorde," *New York Times,* October 24, 2003, as is the quote by Howard Beach resident Betty Braton.

The quote by aviation writer Ron Swanada is drawn from his article "The Need for Speed," *Air and Space Magazine* (March 2004), https://goo.gl/xiQ4A1.

Details about the trial stemming from the crash of Flight 148 and the cause of the crash are drawn from Don Phillips, "Manslaughter Trial of Six Starts in 1992 Air Crash," *New York Times,* May 2, 2006, as is the quote by safety expert H. Clayton Foushee Jr.

Details on the outcome of the trial held in connection with the 1992 crash of Air Inter Flight 148 and the quote by Alvaro Rendon, president of ECHO, are drawn from Nicola Clarke and Don Phillips, "France Clears Six of Charges from 1992 Plane Crash," *New York Times,* November 7, 2006.

Details on the verdict of the 2010 criminal trial stemming from the Concorde crash are drawn from Saskya Vandoorne, CNN, December 6, 2010, https://goo.gl/t6ocyy; and Lucien Libert, "Concorde Trial Starts Ten Years after Crash," Reuters, January 29, 2010.

Quotes by Continental attorney Olivier Metzner and Christian Marty family attorney Roland Rappaport are drawn from Devorah Lauter, "Continental, Mechanic Guilty of Manslaughter in Concorde Crash," *Los Angeles Times,* December 7, 2010.

Details on the French appeals court decision are drawn from Nicola Clarke, "French Court Overturns Convictions in Concorde Crash," *New York Times,* November 29, 2012.

Quotes by French appeals court judge Michele Luga are drawn from Henry Samuel, "Concorde Crash Conviction Overturned by French Court," *Telegraph,* November 29, 2012, as is the quote by victims' rights advocate Stéphane Gicquel.

Details and quotes by attorney Paul LaValle are drawn from the author's phone interview with him, March 8, 2018.

Details and quotes by attorney Gary D. McCallister are drawn from the author's phone interview with him, March 10, 2018.

The personal account of Concorde passenger Trish Dainton of London, recalling the trip she took on Concorde with her late husband, Steve Dainton, as well as the account of amateur photographer Jetinder Sira's experiences photographing

British Airways Concordes, is drawn from "Concorde Stories: Remembering the Pocket Rocket," BBC News, November 22, 2017.

13. A Supersonic Future

The reference to early plans for a Concorde B model to be ready by 1982 is based on information in Glancey, *Concorde*, 233.

The quote by Blake Scholl reflecting on childhood visits to see his grandfather and the distance that now separates his relatives is drawn from Benét J. Wilson, "High Flyer Interview: Blake Scholl, CEO, Boom Technology," Airwaysmag .com, November 15, 2016, https://goo.gl/stFWQb.

The quote by Scholl about lack of progress during 60 years of the jet age and aviation's failure to keep pace with other transportation advances, including the advent of self-driving cars, is drawn from Tamara Chuang, "A Supersonic Jet Faster Than the Concorde Will Get Public Design Debut in Centennial," *Denver Post*, November 14, 2016. Details about Boom Technology's plans to build a supersonic aircraft that will fly 1,450 miles per hour (2,333 kph) and make the transatlantic flight from New York to London in three hours and 45 minutes, as well as the anticipated $5,000 roundtrip ticket price, are drawn from ibid. Background information about Boom, including the year the company was founded and the participation of Scholl's cofounders, Joe Wilding and Josh Krall, is drawn from ibid., as is Scholl's statement about making supersonic flight affordable for families.

The reference to Boom Technology's plans to have its fleet of supersonic planes operating by 2023 is based on information in Jason Steele, "Baby Boom: A Concorde for the 21st Century?" *Air and Space Magazine* (January 2017): 1.

Details regarding the X-B1 demonstrator known as Baby Boom, and plans to manufacture it in North Carolina and test it in Denver and at Edwards Air Force Base, are drawn from Tamara Chuang, "Centennial's Boom Tech Gets a $33 Million Funding Boost for Its Supersonic Jet," *Denver Post*, March 22, 2016. Information about Boom Technology's mock-up of Baby Boom, unveiled in November 2016 at Hangar 14 at Centennial Airport in Denver, is drawn from ibid. and from Jay Bennett, "This Slick Jet Could Repave the Way for Commercial Supersonic Flight," *Popular Mechanics*, November 15, 2016.

Scholl's quote about the failure to take supersonic travel mainstream is drawn from Rob Tracinski, "The Supersonic Age: An Interview with Blake Scholl of Boom Technology," *Real Clear Future*, February 12, 2017, https://goo.gl/ZH2gK6, as is Scholl's quote about political resistance to overland supersonic travel in the United States in the 1970s.

Scholl's quote about his early idea of supersonic flight as the stuff of sci-fi fantasy is drawn from Russell Hotten, "Dubai Airshow: Building a New Supersonic Airliner," BBC News, November 13, 2017.

Scholl's quote about his first startup and his desire to move on to something that would "make a difference in the world" is drawn from Craig Cannon, "Founder Stories: Blake Scholl of Boom Technology," March 24, 2017, https://goo.gl/D9gJ5L.

Scholl's quote about Concorde's "awful" fuel economy and resulting high ticket prices is drawn from Alicia Wallace, "Richard Branson and Boom Technology May Build the Next-Gen Supersonic Plane in Denver," *Denver Post*, March 28, 2016.

Details about a model of Baby Boom being wind tunnel–tested at Wichita State University's National Institute for Aviation Research are drawn from Chuang, "Centennial's Boom."

Details regarding Virgin Galactic and the Spaceship Company's decision to back Boom Technology, provide engineering support, and secure an option to buy 10 of Boom's planes are drawn from Wallace, "Richard Branson and Boom Technology," as is the quote by Richard Aboulafia about the uncertain future of commercial supersonic transport.

Details about Japan Airlines' commitment to invest $10 million in Boom Technology, its option to buy 20 planes, and the aircraft's $200 million price tag are drawn from Ankit Ajmera, "Japan Airlines Invests $10 Million in Supersonic Jet Company Boom," *Reuters Business News*, December 5, 2017, as is Boom Technology's prediction that its plane will produce a sonic boom 30 times quieter than Concorde's.

The reference to action taken by Congress in July 2017 directing the Federal Aviation Administration to revisit the subject of acceptable noise levels generated by sonic booms and the FAA's subsequent review of regulations is based on Kerry Lynch, "Congress Eyes Future of Supersonic Travel," *AINonline*, July 3, 2017, https://goo.gl/mgk28E.

Facts provided regarding the width of a sonic boom cone produced by a supersonic plane flying at 50,000 feet (15,240 m) are drawn from Yvonne Gibbs, "NASA Armstrong Fact Sheet: Sonic Booms," August 15, 2017, https://goo.gl/HBZHXw.

Facts concerning the pressure (1.94 psf) generated by a Concorde boom over land, producing a sound experienced as 105 dB(A) on the human ear, are drawn from Eli Dourado and Samuel Hammond, *Make America Boom Again: How to Bring Back Supersonic Transport*, Report, Mercatus Center, George Mason University, October 27, 2016: "Comparisons to Noise Levels We Already Accept," 25.

The reference to a Concorde traveling at 50,000 feet (15,240 m) exceeding noise levels of 110 decibels is based on reporting by Robert Silk, "Lawmakers Want FAA to Review Rules for Supersonic Aircraft," *Travel Weekly*, August 15, 2017, https://goo.gl/N4Faap; "Noise of Concorde Intolerable," BBC News, August 2004; and Dourado and Hammond, "Make America Boom Again," 3.

Details about Operation Bongo and the public reaction to daily supersonic flights over Oklahoma City in 1964 are drawn from Ken Raymond, "Jets Broke Speed Barriers and Tempers," *Oklahoman*, September 18, 2006, https://goo.gl/MhafSR, as are quotes by Bridget Meadows regarding her and her children's response to sonic booms during Operation Bongo.

Interpretive information regarding the effects on the human ear of 110 decibels and comparable noises is drawn from Purdue University Chemistry Department, "Noise Sources and Their Effects," February 2000, https://goo.gl/5QRVTt.

The reference to the estimated decibel range of a typical Concorde sonic boom is based in part on data provided by Dourado and Hammond, "Make America Boom Again"; see also "Noise of Concorde Intolerable," BBC News, August 2004.

Information about NASA's Low Boom Flight Demonstrator X-plane and its Quiet Supersonic Technology program and partnership with Lockheed Martin, as well as the $20 million contract awarded in February 2016 to Lockheed Martin and its subcontractors, is drawn from Sarah Ramsey, "NASA Begins Work to Build a Quieter Supersonic Passenger Plane," NASA press release, February 29, 2016, https://goo.gl/P9hbXN.

Quotes by NASA manager John Carter about the X-plane and the next generation of supersonic planes are drawn from Robert Silk, "Big Boom or Big Bust," *Travel Weekly*, January 11, 2017, https://goo.gl/R7eZcQ.

Details about Aerion Corp. and its AS2 business jet, including its passenger capacity, wing design, and projected speed, are drawn from Christian Davenport, "Lockheed Martin Teams Up to Build Supersonic Business Jet," *Washington Post*, December 16, 2007; see also Aerion Corp., "Aerion and Lockheed Martin Join Forces to Develop the AS2," press release, December 15, 2017, https://goo.gl/yg4bDw.

Details about Spike Aerospace, the S-512, and plans to produce a low-boom supersonic business jet are drawn from Jay Bennett, "Spike Aerospace to Fly Scaled Demonstrator of Supersonic Jet," *Popular Mechanics*, September 25, 2017, https://goo.gl/wQKxia; and from Spike Aerospace, "The Spike S-512 Supersonic Jet," https://goo.gl/tMNzn5.

Quotes by Thomas Corke concerning the features of low-boom supersonic aircraft are drawn from Lockheed Martin, "Quick and Quiet: Supersonic Flight Promises to Hush the Sonic Boom," https://goo.gl/ry3nbn.

The reference to Boeing's plans to develop a hypersonic spy plane is based on reporting by Guy Norris, "Boeing Unveils Hypersonic 'Son-of-Blackbird' Contender," *Aviation Week*, January 11, 2018, https://goo.gl/aaF51R.

The quote by Boeing executive Dennis Muilenburg is drawn from Phil LeBeau, "Boeing Planning on Hypersonic Jets for Commercial Flights, Though the Concorde's Memory Lingers," CNBC, June 19, 2017, https://goo.gl/bMi6gT, as are details about the speed of hypersonic jets and projected flight times.

Details about the ceremony on the 10th anniversary of the Concorde crash are drawn from Jeffrey Schaeffer, "Families Mark Ten Years Since Concorde Crash," *San Diego Union-Tribune*, July 25, 2010; see also "Crash of Concorde, 10 Years Ago Today," *L'Express and AFP*, July 25, 2010.

The quote by Claudine Le Gouadec, sister of Concorde flight attendant Virginie Le Gouadec, is drawn from Schaeffer, "Families Mark Ten Years," as is the quote by Gonesse resident Claude Philippe.

The description of the Concorde monuments in Gonesse and near Charles de Gaulle Airport is based in large part upon details provided by the FAA Lessons Learned website, "Accident Memorial," https://goo.gl/BHrWcA.

Index

Page numbers in italics indicate photographs and illustrations.